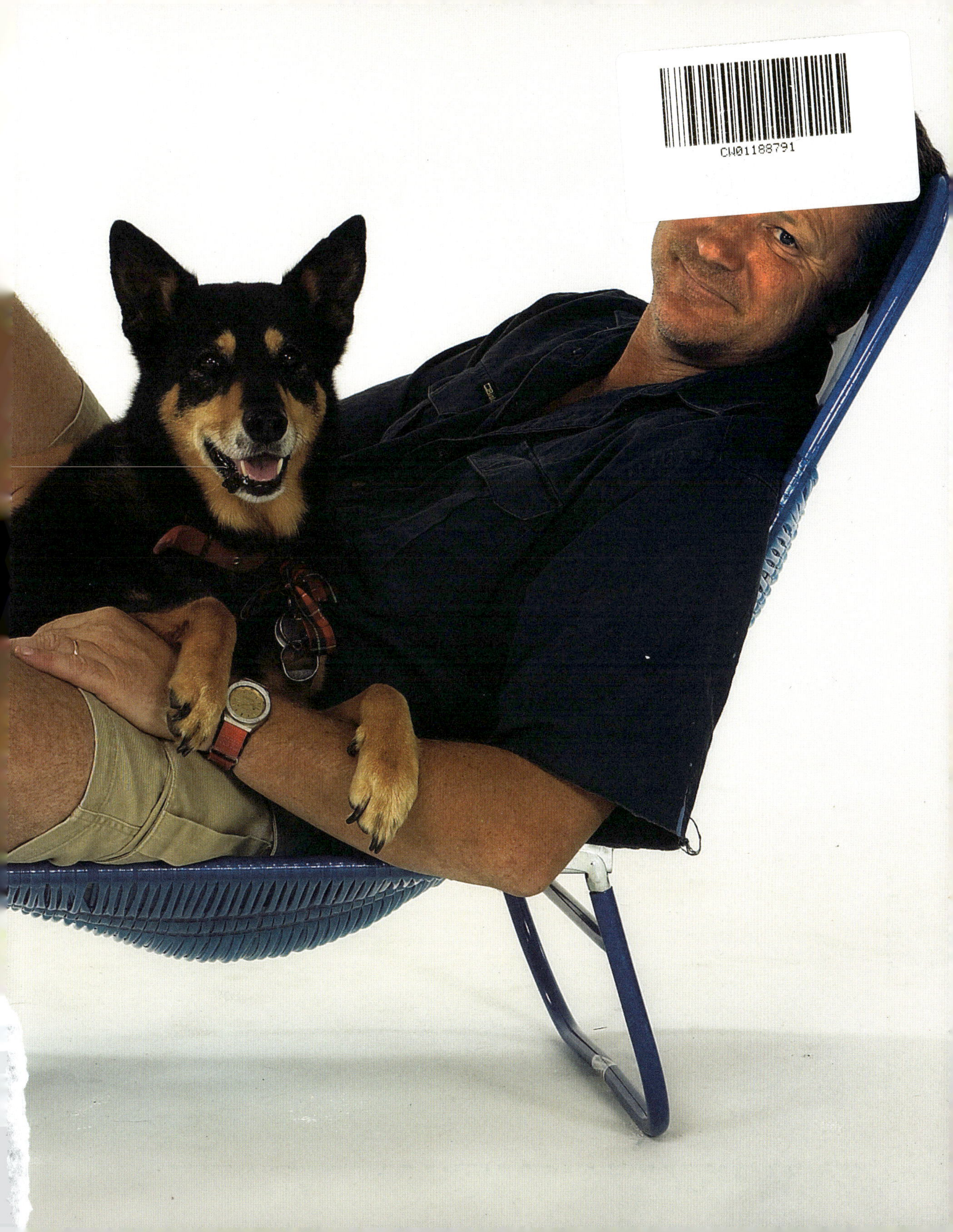

Out the Back

Scott Cam

Published by Murdoch Books®, a division of Murdoch Magazines Pty Ltd.
Murdoch Books® Australia, Pier 8/9, 23 Hickson Rd, Sydney NSW 2000
Phone: 61 (02) 4352 7000   Fax: 61 (02) 4352 7026
Murdoch Books UK Ltd, Erico House, 6th Floor North, 93-99 Upper Richmond Road,
Putney, London SW15 2TG
Phone: +44 (0) 20 8785 5995   Fax: +44 (0) 20 8785 5985

Chief Executive: Juliet Rogers
Publisher: Kay Scarlett
Design concept: Marylouise Brammer
Designer: Tracy Loughlin
Photographer: Suzie Mitchell
Project Manager: Sarah Baker
Editors: Roland Arvidssen and Sarah Baker
Illustrator: Genevieve Huard
Production: Monika Vidovic
Editorial director: Diana Hill

Printed by Midas Printing (Asia) Ltd. Printed in China.
National Library of Australia Cataloguing-in-Publication Data:
Cam, Scott. Out the Back. Includes index.
ISBN 1 74045 391 3. 1. Garden structures - Design and construction -
Amateurs' manuals. 2. Family - Anecdotes. I. Title.
717

First printed 2004. Text © Scott Cam 2004. Photography and illustrations © Murdoch Books. All rights reserved. No part of this publication may be reproduced, stored in any retrieval system or transmitted in any form or by any means, electronic, mechanical, photocopying, recording or otherwise without the prior written permission of the publisher. Murdoch Books® is a trademark of Murdoch Magazines Pty Ltd.

Out the Back

Scott Cam

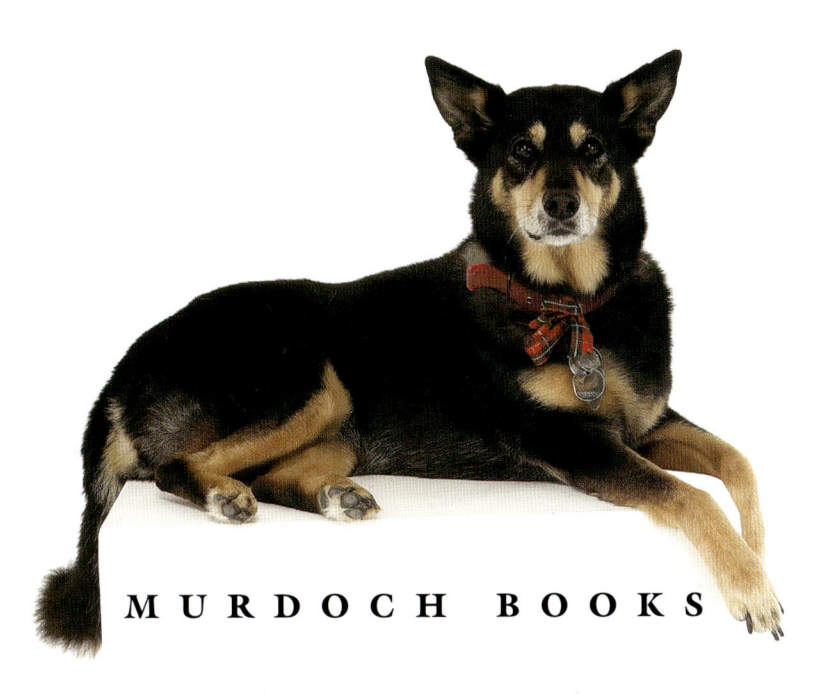

MURDOCH BOOKS

Contents

Introduction	6
Backyard projects	14
Backyard maintenance	84
Enjoying the backyard	122
Domestic duties	162
Table of knowledge	176
Index	190

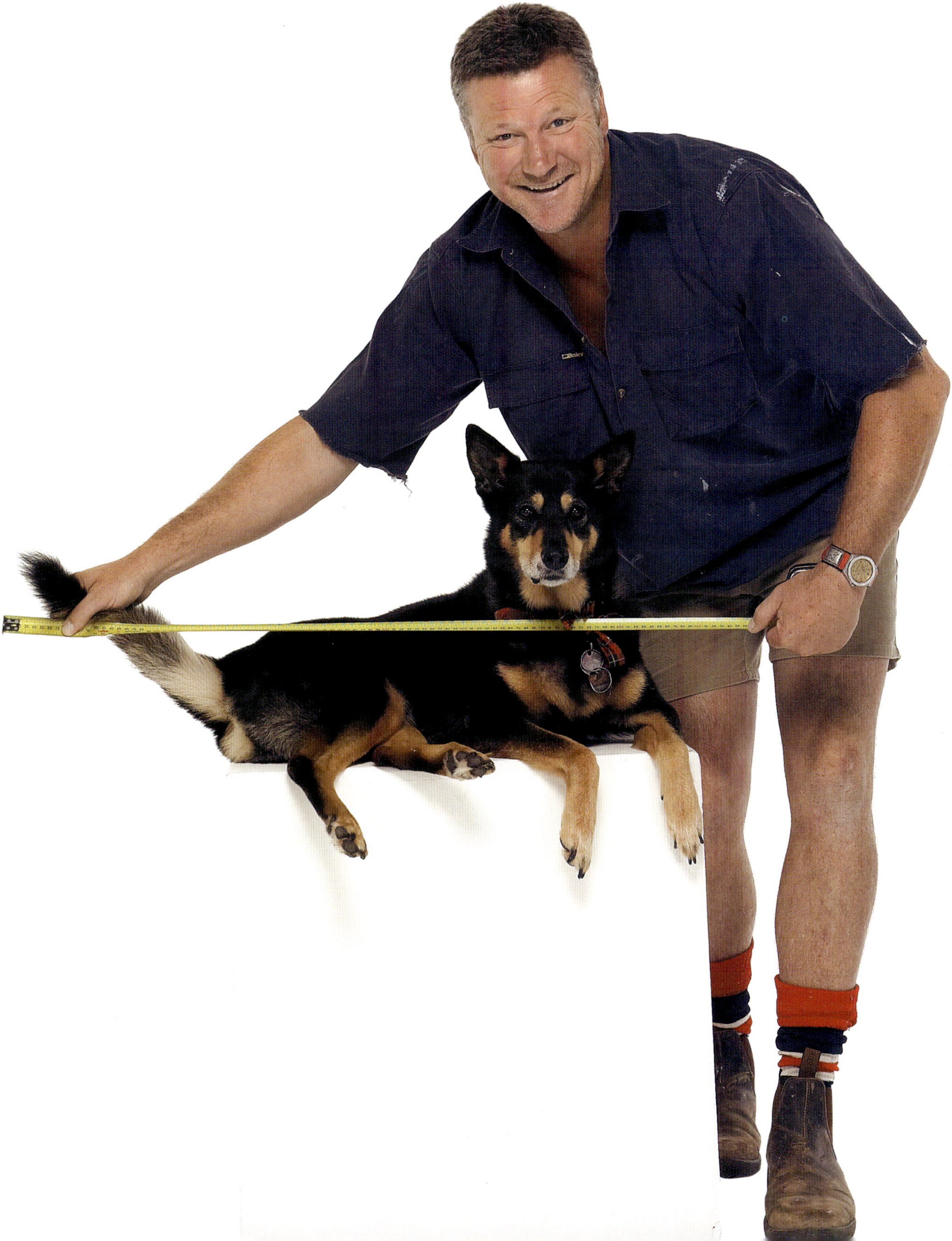

Introduction

This book is about how to be a bloke, plain and simple. Back to basics, easy to follow.

Just like Germaine Greer campaigning for feminists in the 1960s, I'm doing the same thing for us downtrodden blokes.

We're not allowed to do anything blokey anymore. It's too 'yobbo', or too 'insensitive'.

Well, it's time for a revolution.

And if anyone asks what you're doing, tell 'em you're gonna build a fire, cook some snags, crack open a can and watch the footy on telly. And when they object, interrupt them with, 'Get us a stubby holder, would you...please!'

So this book has just a few tips on the dying art of being a bloke. You don't have to be a chauvinist or insensitive or messy or rude, just a good old honest Aussie Bloke.

I'll tell you how to build a decent fire and cook snags to perfection, run a keg with cold beer and a decent head, build a few things in the backyard for the Missus and the kids, and even how to do a few jobs inside the house.

Basically, I want you outside, being useful and looking the part.

Knock the snag out of them early

I reckon the only way to get on top of this situation is to look to the future. The snags (sensitive New Age guys) out there at the moment...well, they're already gone, their spirits broken, just shells of the blokes they once were.

I'm not forgetting about these blokes, it's just that they need a lot of work, whereas the young blokes only need a little bit of moulding. Before you know it, you've knocked the snag out of them early.

My older boy is seven years old, and he already has his own

shed with a small bench, small vice and a good set of tools. Charlie keeps his pushbike and skateboards in his shed, and he hangs around in there knocking stuff up. I'm so proud.

In winter Charlie builds the fire in the lounge room. His Mum helps him light it, and he stokes it and keeps it going until I get home. What a little champion.

I don't reckon my two boys'll be having facials when they grow up.

What is a blokosexual?

The blokosexual is almost the complete opposite of the metrosexual, although to be honest I'm not entirely sure what a metrosexual is or does, so I'd best just talk about the blokosexual.

I'm a self-confessed blokosexual. My idea of a great day is out in the backyard with the family barby going, esky full of essentials, maybe a few friends. Telling yarns, laughing and drinking cold ones, eating snags and salad with iceberg lettuce. (What's the go with all this chook food they're passing off as lettuce? Most of it tastes like a gum leaf. Iceberg is the lettuce of choice for the blokosexual.)

Sport plays a big part in the blokosexual's life. Participation is not essential — watching on the tube or at the ground is quite acceptable.

Having said that, gone are the days of the old armchair commentator. The blokosexual stays reasonably fit — maybe a bit of a light run with the boys before the pub. That way you can pop in and have a couple of beers and still stay in good nick. That's the other thing — the blokosexual likes to live a long time.

The preferred holiday is camping in the bush. I like to camp in the swag —

not a tent. The Bride's not overly keen on that one, so sometimes the tent comes out.

Winter time sees the blokosexual move straight into fire mode — cooking on the open-fire barby. There's nothing more Aussie than burning some chops and snags on the open-fire barby. And I like to keep the house warm with the open fire inside. I love that smell of fire in the house the next morning.

That's a rough outline of the blokosexual's movements. But, more importantly, the things he doesn't do are just as important.

Facials are out. Having a facial would be like breaking all Ten Commandments on the same day. More than one clothes shopping day a year is also a no-no. And what's wrong with a good old-fashioned barber? Or you could go to the extreme and cut your own hair like I do. I've been cutting my own hair for years. I've got my own barbershop at home. It's an old mirror nailed up on the back fence. I run a lead out from the garage for the clippers and, hey presto, I call it Number Twos. There are no queues, and it's cheap as chips.

Anyway, by now I think everyone knows what a blokosexual's all about.

He likes his sport, he doesn't mind the odd beer. He keeps reasonably fit, and he works hard. He's not overly concerned about his appearance, but he is not a slob. And he is a true-blue, dinky-di Aussie.

Backyard projects

Nobody's impressed with a bloke who started a deck and never finished it. So don't bite off more than you can chew, but when you do take a bite, chew like buggery! I reckon the best project in the backyard is snags on the barby, friends and cold beers. No one can stuff that up. Good luck!

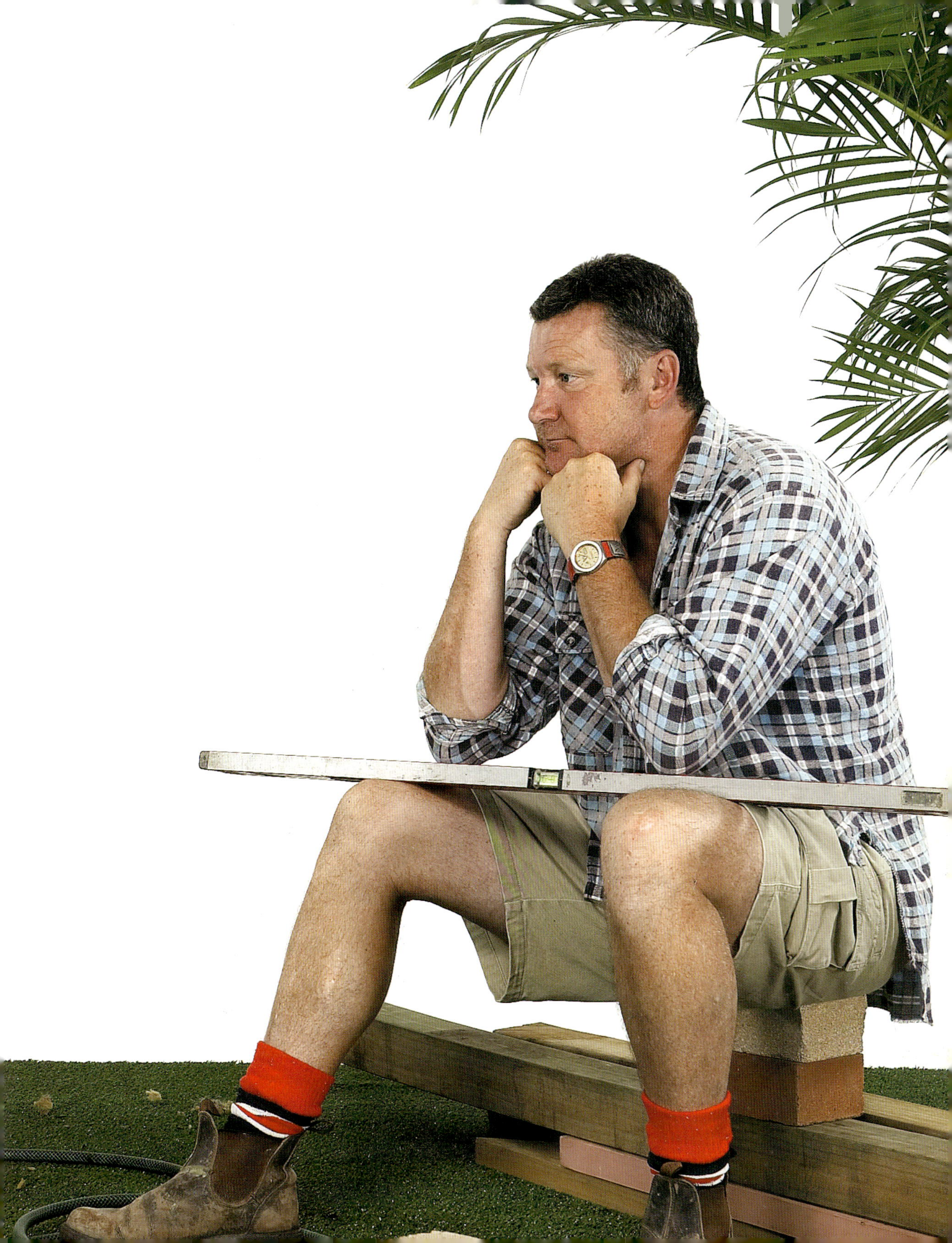

DO YOUR HOMEWORK!

OK, so you've ditched the sensitive New Age approach and now you're in touch with your inner bloke. Good stuff. You've got all these grand plans for your backyard, you're busting with enthusiasm...but just relax for a second.

Before you end up with your delicates in a twist, you need to get your head around a couple of basic skills. In just a few pages, you'll know how to set up string lines, put a post in the ground, choose the right timber and get it level. These are all great ways to avoid stuffing up that deck or pergola you're planning to build. Best of all, you'll be able to speak fluent tradesmanese with the hardware people, preferably in front of that mate who's agreed to give you a hand over the weekend.

Setting up a string line

If you're going to be putting up any kind of structure in the backyard, getting it straight is the most important part of the job. Unless you want a pergola that looks like the Ettamogah Pub.

string lines are the way to go. It's pretty simple to set one up: whack a couple of garden stakes into the ground, and stretch a piece of string between them. You can get specialized spools of the stuff from your local hardware.

When you're building something attached to your home, run your main string line down the side of the house — that way it's as close to being square with the walls as it can be. Use this line as a reference point when setting up your other lines.

Now for the tricky part. There's no point setting up string lines unless they're in the right place. If you need the string line to be level — and you often do — you pop what's called a line level onto it. This is like a baby spirit level with tiny little hook things that hang off the string line.

Getting the corners square is your next mission. To get a right angle with string lines, use what's called the 3:4:5 method.

Measure 3 m (or 3 yd) down one string line and 4 m (4 yd) down the other, and mark both those points. You'll need a friend to get this right. Now measure the distance between the two points. This line is known as the hypotenuse (in other words, the diagonal of the triangle). If the two string lines are exactly at right angles, that distance should be exactly 5 m (5 yd).

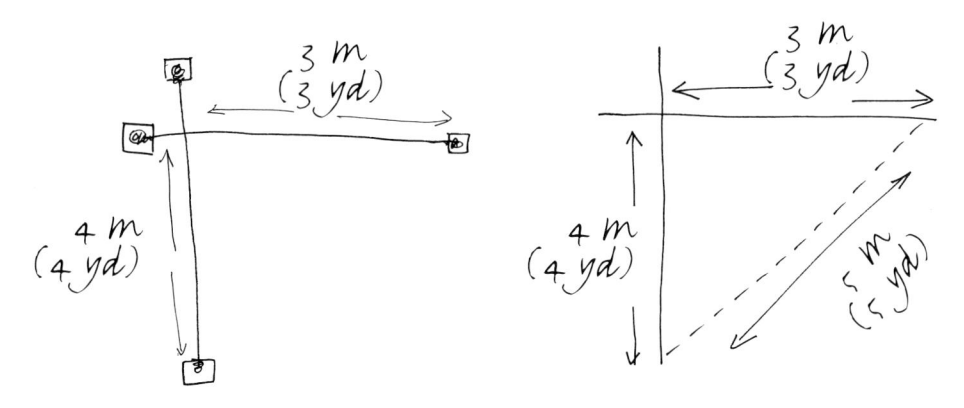

Because the main line always stays, the line with the 3 m (3 yd) measurement gets adjusted to create the 5 m (5 yd) hypotenuse, which in turn will give you the perfect right angle.

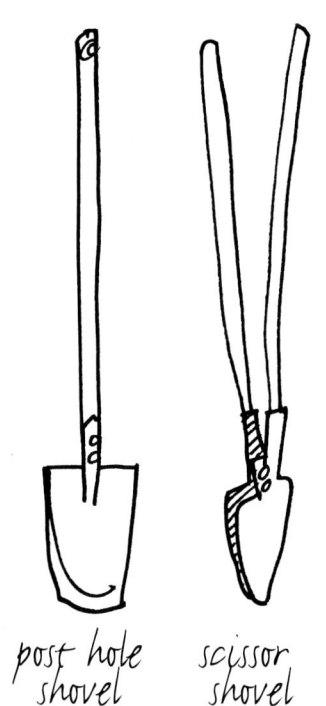

Cementing in a post

The easiest way to set a post in the ground in the backyard – a post for a fence, deck, letterbox, whatever – is to use quick-setting concrete.

Whenever you put any post in the ground, it should be at a depth of 600 mm (2 ft). The exception to this rule might be a letterbox post, where 400–500 mm (16–20 in) would be acceptable.

post hole shovel

scissor shovel

You'll need a scissor shovel or post hole shovel, level, and a hose or bucket.

If you have neither of these shovels, any spade will do — it's just that the scissor shovel makes life so easy.

The best timber to use in an in-ground situation is H4 treated pine. (See page 21.) The minimum post is 90 × 90 mm (3½ × 3½ in) — H4 treated pine. If you have a specific situation — such as a fence post — then of course you can use a 125 mm × 50 mm (5 × 2 in) morticed fence post; otherwise stick to 90 × 90 mm.

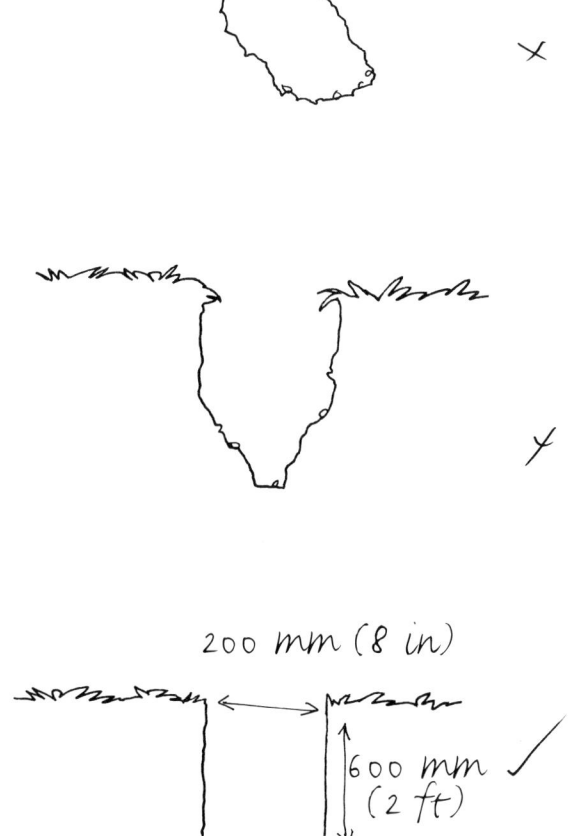

Dig your hole with about 50 mm (2 in) extra all round your post so that your hole for a 90 × 90 mm post is about 600 mm (2 ft) deep × 200 mm (8 in) square. Make sure that the hole is straight and doesn't taper in at the bottom.

Drop your post in the hole. If you are using quick-setting concrete, fill a third of your hole with water. Have a bag of quick-setting concrete with one end torn open at the edge of the hole.

TIP OF THE DAY:
CONCRETE BAGS
There are two types of concrete or cement bags — 40 kg (90 lb) and 20 kg (45 lb). Don't be a hero and muck around with the 40 kg (90 lb) bags. The 20 kg (45 lb) bags are much easier to handle and great on the back. I always use 20 kg (45 lb). If I come across a 40 kg (90 lb) bag, I cut it in half with a Stanley knife.

It's always handy to have a mate to give you a hand — it's just a matter of one person levelling on both sides of the posts and the other pouring the concrete in. If you are on your own it is the same process, just a bit of a battle.

Once the concrete is in, use a stick to poke and prod it. This will bring a little moisture to the top and mix it up a bit, making for a more consistent mix. While you are doing this, keep checking your levels, and check them one more time as you walk away from the finished post. It is much easier to adjust now.

You will always end up with cement dust on the post: just give the bottom of the post a quick hose and you are done.

TIP OF THE DAY: 'NORMAL' CONCRETE

IF YOU'RE USING NORMAL CONCRETE, YOU MUST MIX IT UP WITH WATER IN A BUCKET OR WHEELBARROW AND POUR IT INTO THE HOLE WET.

What timber to use where

Treated pine (H4, H3)

The code indicates a hazard rating, as the timber has been treated with preservative for specific purposes. Use H3 or H4 as external timbers.

* Use H4 where the timber will be in contact with the ground – for example, for posts.

* Use H3 for above-ground use – for bearers, joists, decking boards and pergola rafters.

TIP OF THE DAY:

GALVANIZED FITTINGS
Always use galvanized or stainless steel bolts, fittings or fasteners with treated pine. Never use bright steel or zincalume.

Pink-primed timber (H3)

This is pine that has been coated with a primer, which is normally laminated. The advantages are that it is always straight, and it comes in a variety of sizes and in much longer lengths than you can get with other timber.

Pink-primed timber has a hazard rating of H3, so it's for above-ground use only. Perfect for handrails and pergolas. It is a little bit more expensive but it looks good and paints up terrifically.

TIP OF THE DAY:
CUT ENDS
Whenever you cut treated timber, you must treat the cut ends with an appropriate product. Always wash up well after using treated timber.

Anyway, I could go on about different kinds of timber for hours – in fact, I will a bit later.

Water level

This is a cheap, accurate and essential tool for the handyman. Basically, a water level is just a length of clear hose filled with water that enables you to transfer a measurement from one point to another. So, for example, when you're setting out the posts for a deck, you need to make sure that you mark each post to the same height so your decking boards will be completely level. The water level is a dead easy way to do that.

The water level can be any length, so whip down to the hardware and buy about 5–10 m (16–32 ft) of clear 10–15 mm (3/8–5/8 in) hose. The great thing is, it doesn't matter what size it is. Mine is a 12 mm (1/2 in) hose about 10 m (16 ft) long. This size of clear hose is cheap as.

The great thing about water is that it finds its own level. If you half fill a glass of water and tip it at an angle, the water will stay level.

So, fill your clear hose with water by sticking one end in the hose tap out the back and leaving the other end on the ground. Put stoppers in either end of the hose so the water doesn't run out. Make sure you can see all the hose, because you must remove all the air bubbles. This could take a couple of goes. Take your time and do it right.

Like the water in the glass, the water in the tube will find its level no matter where the ends are.

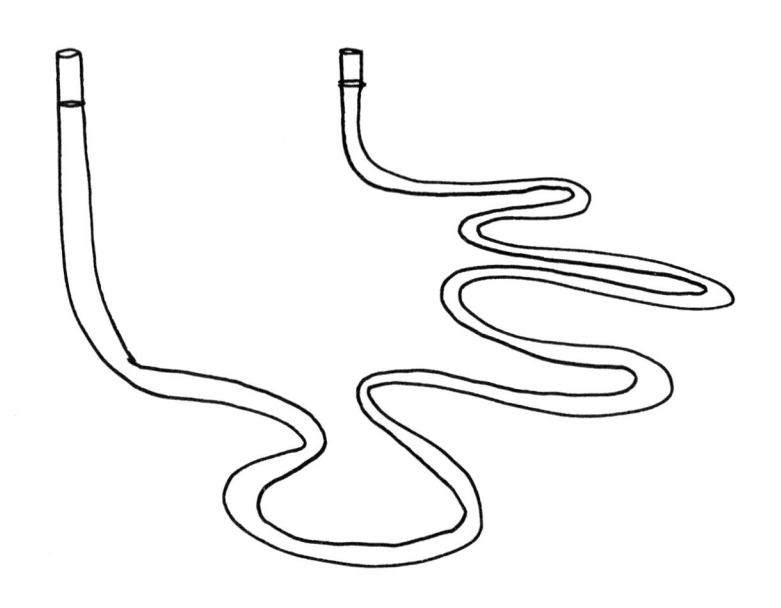

So, when you want to make sure your posts are level, you need to transfer a mark on one post to the other posts. You'll need a friend to help with this part. Take the stoppers out of the ends while you're using it, and just use your thumbs when you need to stop water running out. Otherwise the air pressure in the hose will mess with the system or create bubbles. And you won't get an accurate level unless the water is moving freely. Hold one end of the water level on the existing mark, then get your mate to take the other end of the hose to where you want the new mark. Theoretically, it could be 2 km (1¼ mi) away if your hose is long enough.

Tell your mate to move his end of the hose up or down against his post until the water line stops on your mark. When there is no movement, tell your mate to mark his water line. Both marks will be perfectly level.

Now that you've bravely struggled through all the background stuff, it's time to build a project that will really impress those who count — the kids and the Minister for Jobs around the House, or maybe your mates, or your Mum, or what about your husband? Whoever it is, if you pull this one off, those you most want to impress will be standing on your deck in disbelief, because they wouldn't have a clue how to start something like this.

Well, here's how to start.

THE POSTS AND BEARERS

THE FIRST THING TO DO IS THE SET-OUT, THAT IS, ESTABLISH THE SHAPE AND SIZE OF WHAT YOU'RE BUILDING. I'VE SAID IT BEFORE AND I'LL SAY IT AGAIN: THIS IS THE MOST IMPORTANT PART OF THE JOB. IF THE SET-OUT IS RIGHT, THE FINISHED PRODUCT WILL LOOK RIGHT.

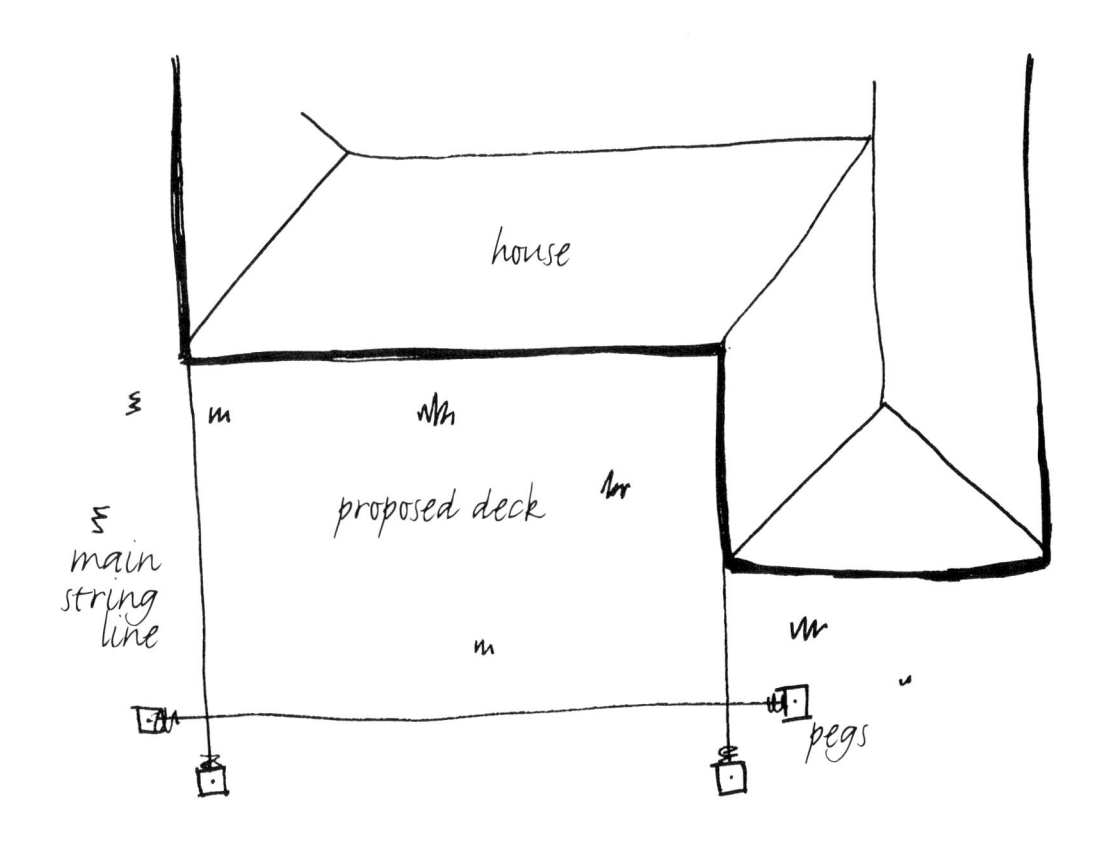

The set-out is based on something you already have in the backyard, be it the house or the garage. Generally, the set-out will come square off whatever your deck is attached to.

Set up a string line along the side of the house. Measure off this line to get the position of your other lines and use the 3:4:5 method to get your right angles...well, right (see page 17).

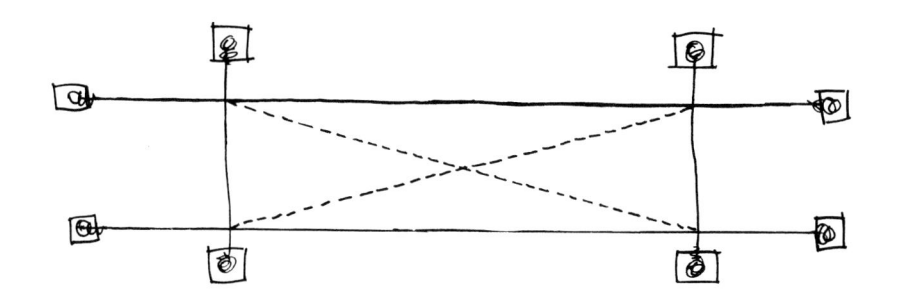

Measure the diagonals from one corner of your deck set-out to the other. If the diagonals are of equal length, the corners are square.

Now you have a set of string lines representing your deck, set the posts in the ground with quick-setting concrete (see page 18). It's important to space your posts the correct distance apart. Timber will span a certain distance before it starts to bow. Check with your timber yard and they'll tell you what spans to go with. There are Australian Standards for all sizes of timber.

I always build a low deck with 90 x 90 mm (3½ x 3½ in) posts, 150 x 50 mm (6 x 2 in) bearers, 100 x 50 mm (4 x 2 in) joists, and 90 x 20 mm (3½ x ¾ in) deck boards (all treated pine) or hardwood.

DETERMINE WHICH WAY YOUR BEARERS ARE GOING TO RUN — YOUR DECKING BOARDS WILL END UP RUNNING IN THE SAME DIRECTION.

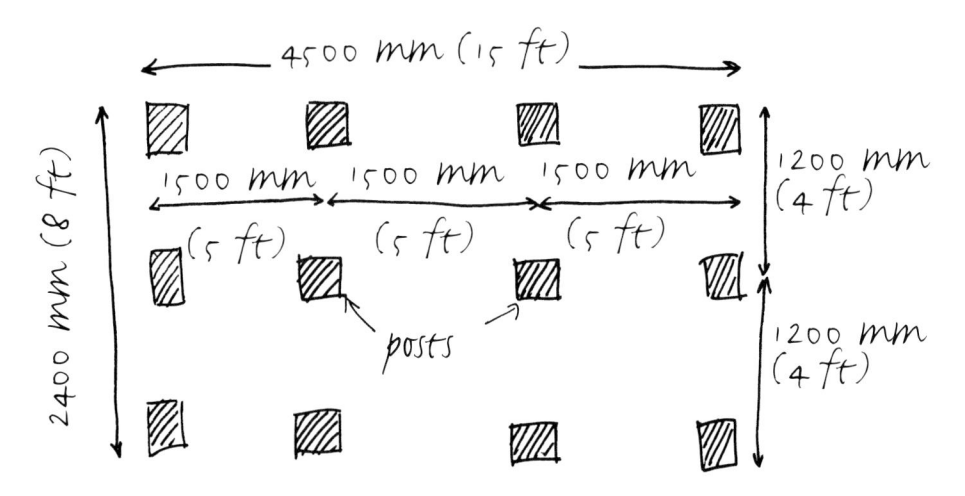

[Diagram: A rectangle measuring 4500 mm (15 ft) wide by 2400 mm (8 ft) tall, with three rows of posts. Posts along the top row are spaced 1500 mm (5 ft) apart. The middle row of posts is 1200 mm (4 ft) down from the top, and the bottom row is another 1200 mm (4 ft) below that.]

THE BEST WAY TO SET UP A BEARER POST RUN IS TO DO THE OUTSIDE POSTS FIRST, THEN STRING BETWEEN THEM SO THEY ALL LINE UP PERFECTLY.

I CENTRE THE POSTS ABOUT 1500 MM (5 FT) APART USING 150 MM x 50 MM BEARERS. THE JOISTS, 100 x 50 MM (4 x 2 IN), WILL SIT ON THESE BEARERS, WHICH CAN BE CENTRED ABOUT 1200 MM (4 FT) APART. IN OTHER WORDS, IF YOUR BEARERS ARE RUNNING LEFT TO RIGHT LIKE IN THE PICTURE OVERLEAF, THERE SHOULD BE NO MORE THAN ABOUT 1500 MM (5 FT) BETWEEN THE CENTRE OF A POST AND THE CENTRE OF THE ONE BESIDE IT, AND 1200 MM (4 FT) IN THE OTHER DIRECTION.

post holes

posts

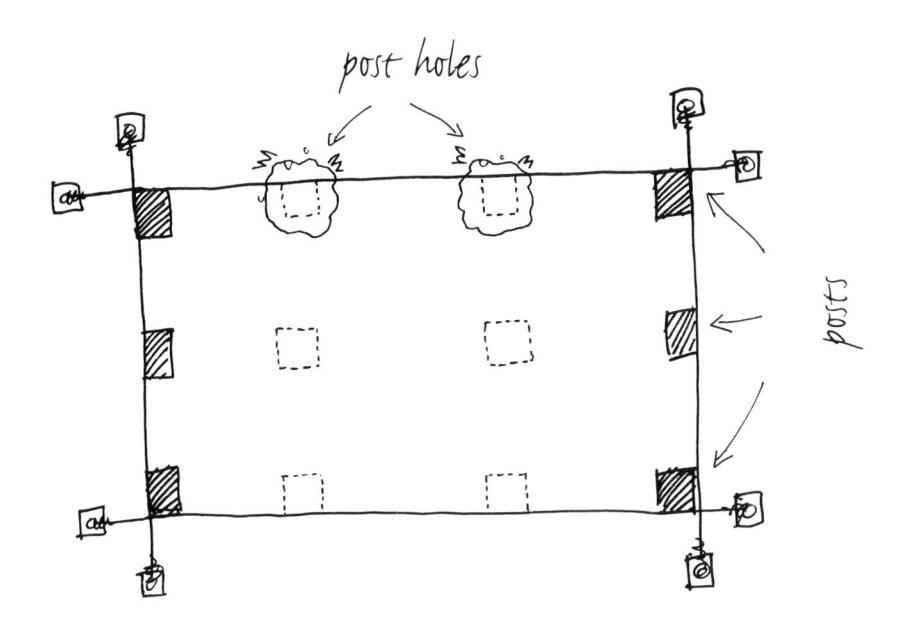

TIP OF THE DAY: BEARER HEIGHT

TO GET THE HEIGHT OF YOUR BEARER, DETERMINE THE TOP OF YOUR DECK. USUALLY IF YOU'RE COMING OFF THE HOUSE IT'LL BE THE UNDERSIDE OF THE DOOR SILL. TRANSFER THAT POINT ACROSS TO THE NEAREST POST USING A SPIRIT LEVEL — THEN MEASURE DOWN 20 MM (3/4 IN) FOR DECKING — AND 100 MM (4 IN) FOR JOISTS.

DOUBLE-CHECK YOUR TIMBER SIZES TO MAKE SURE YOU MEASURE DOWN CORRECTLY. THEN MARK 120 MM (4 3/4 IN) DOWN AND USE YOUR WATER LEVEL (SEE PAGE 23) TO TRANSFER that mark to every post. Fasten each bearer with two galvanized bolts through each post.

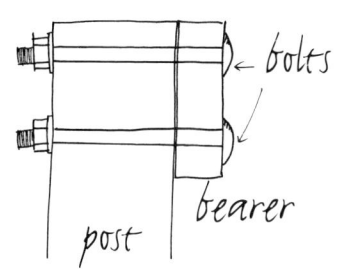

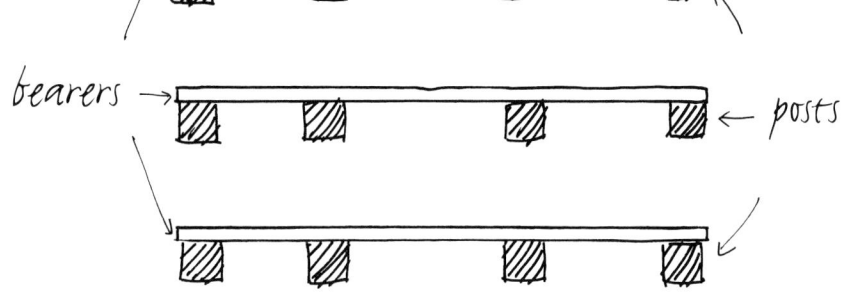

I always leave my posts about 100 mm (4 in) higher than needed and cut them off later. It's better to have too much than too little.

The joists and decking

You are now ready for the joists. Lay 'em across the bearers. They need to be 450 mm (18 in) apart from centre to centre.

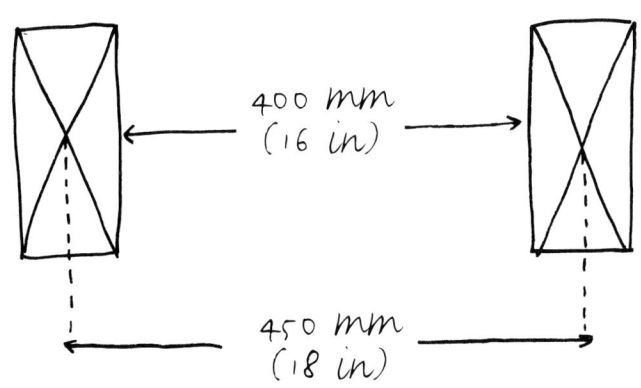

The easiest way to get the gap nice and consistent is to cut a block of timber 400 mm (16 in) long and use it as a spacer between each pair of joists.

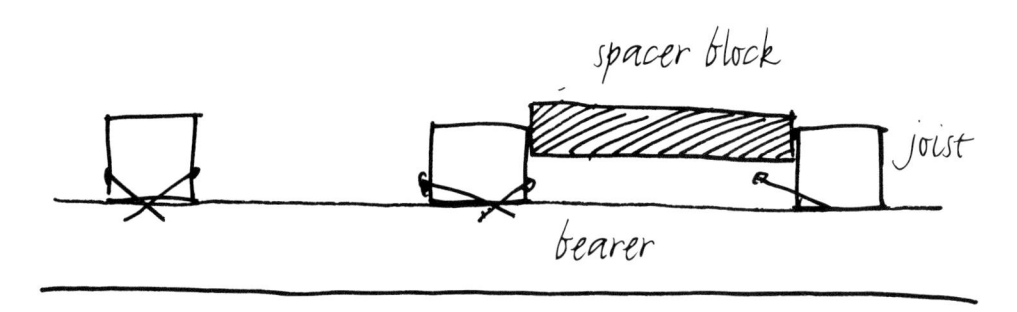

spacer block

joist

bearer

Nail the end joist onto the bearers by skewing galvanized 75 mm (3 in) bullet-head nails in through the side and into the bearer. Once the first joist is in, use your 400 mm (16 in) spacer block to line up the other joists and nail them in place.

OK, when the joists are done it's time to get stuck into the decking boards.

When nailing down decking boards, you have to use galvanized twisted shank decking nails. Absolutely nothing else will do the job.

As a spacer for the boards use a couple of 75 mm (3 in) galvanized bullet-head nails. They're just a bee's dick wider than 3 mm (1/8 in) — the ideal gap between decking boards. Hammer each nail through a small piece of timber to make it easier to handle.

TIP OF THE DAY: Ripple side down

Some boards come with a ripple face, which is at times called a grip deck. In my opinion that's crap. These ripples were originally designed to give air flow at the point where the decking board meets the joist so as to prevent rotting.

If you lay the decking boards ripple side up, they'll gather dirt and hold moisture. The ripples can also splinter up and become dangerous, and they are very uncomfortable under bare feet (trust me). So, ripple side down!

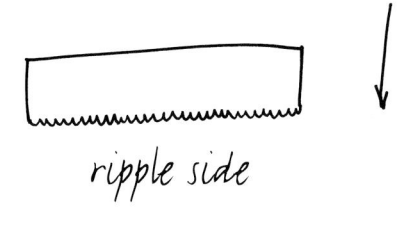

ripple side

Whack in two nails per board at each joist. Be careful not to bruise the timber when nailing home! Treated pine is soft and bruises easily.

You can either pre-cut each board and lay them individually or have a go at my preferred method. Lay all the boards a little bit long and when you are finished, use a nice big straight edge or string line and cut them all at once with a circular saw or hand saw. This makes life a lot easier.

Once you have finished, treat your boards with an appropriate wood preservative. If you do this regularly, your deck will enjoy a much longer life.

Hardwoods

Now you've got that deck finished, crack open a well-earned cold one and I'll give you that yarn on timber that I promised you earlier.

You see, I absolutely love Australian hardwood. Blue gum, spotted gum, tallowwood, blackbutt and jarrah are all terrific timbers.

Used outside and untreated, hardwood tends to weather quite dramatically with UV and moisture. That is why I like to feature these timbers internally, and leave the treated timber for the outside.

That's not to say you can't use these timbers outside — just keep the appropriate oils up to them and keep them reasonably dry. If timber sits in constant moisture, it will rot, no matter what you do.

Oregon

Twenty years ago, it was very common to use oregon in pergolas, decks and for a lot of other external features, as well as for framing timber when building homes.

It really is an internal timber and should never be used to build a deck or pergola. If you ever see a pergola that is half falling down or rotted out, I guarantee it will be oregon. I have rebuilt hundreds of them over the years, so keep the oregon indoors.

TIP OF THE DAY: DRESSED OREGON

OREGON IS USUALLY SOLD AS ROUGH-SAWN. THAT MEANS WITH FURRY BITS ON IT. YOU CAN GET IT DRESSED ALL ROUND — IN OTHER WORDS, SMOOTHED OFF AND WITH A COAT OF OIL ON IT. IT LOOKS FANTASTIC INSIDE!

The ute

It must be a horrible feeling just driving around in a normal sedan – not a ute, I mean. Even worse if your sedan is a hatch or a four-cylinder: not being able to park in loading zones, having to indicate to change lanes. I try not to get into the Minister for Transport's school bus – it's not a good experience for me.

I've got the solution – buy an old ute. You could say you've bought it for the eldest child for when they get their driver's license. It's going to be very handy: taking rubbish to the tip, picking up new purchases like furniture and whitegoods (you'll save on delivery). It'll pay for itself in no time. And when someone is moving house, you could offer to help...because you have a ute.

When you're doing any work around the house, whip down the hardware in your ute, pop around to a mate's place in the ute. It'll boost your credibility.

And when you finally get your ute, you must have a sign painted by a professional. I've always wanted a ute with writing. I saw a bloke down the tip years ago. He had the old Falcon ute, and it had this in sign writing on the side:

BILLY JOHNSTON

Ute owner extraordinaire

Most jobs too big

Mobile 0412 345 678

Brilliant. He probably only brought it out on weekends — parked it out the front of the pub for one on the way home.

If you do it right, you'll get the respect you've always craved. A dog in the back, maybe even 1/4 ton of sand and half a dozen bricks...always nice touches. And on the passenger seat, dozens of empty drink containers — maybe even empty takeaway packets, dried mud from the boots all over the floor.

If you own a ute you've got a big responsibility, mostly to other ute owners: you can't let the team down. Dress correctly, drive with the chest out proud and puffed up. Safely of course, and none of that road rage business (when I see road rage I feel like ramming the offender).

If you've got the flash car on the inside lane and you need to change lanes, just put the indicator on and look in the mirror. It's like Moses parting the Red Sea — nine times out of 10 the flash car or any car that's not a ute will break and let you in. The bloke in the flash car sees you've got 1/4 ton sand, some bricks, maybe an old Esky screwed down to the tray as the smoko box (a great look). He thinks, 'This guy has got to get to work and slog it out in the sun, the poor bugger. I might let him in.'

One final note about owning a ute: when you pass other tradesmen in traffic, wave to them by raising one finger off the steering wheel and they will return the gesture.

How good is that?

THE PERGOLA

The pergola can range from something resembling the Sydney Harbour Bridge right down to something anyone could have a go at. Guess which one I'm going to show you?

Obviously, the shape of this pergola can be changed, but the principles stay the same. Just remember, if you make your design bigger, you must check your timber sizes to make sure they will span the distance you're intending. This pergola design is a simple square one.

Materials
Two 90 x 90 x 3300 mm (3½ x 3½ in x 11 ft) posts
Quick-setting concrete
One 150 x 50 x 2400 mm (6 x 2 in x 8 ft) wall plate
Dynabolts
One 150 x 50 x 3000 mm (6 x 2 in x 10 ft) decorative beam
Two 150 x 50 x 2400 mm (6 x 2 in x 8 ft) rafters
Three 150 x 50 x 2350 mm (6 x 2 in x 7 ft 10 in) rafters

1 Start off by setting out a square, which will represent your pergola. This pergola is coming off the house at 2400 x 2400 mm (8 x 8 ft), enough room for a table and chairs. For a reference point, use an already existing structure, usually the house. Have a quick check over 'Do your homework' (page 16) if you need a little reminder on how to do the set-out.

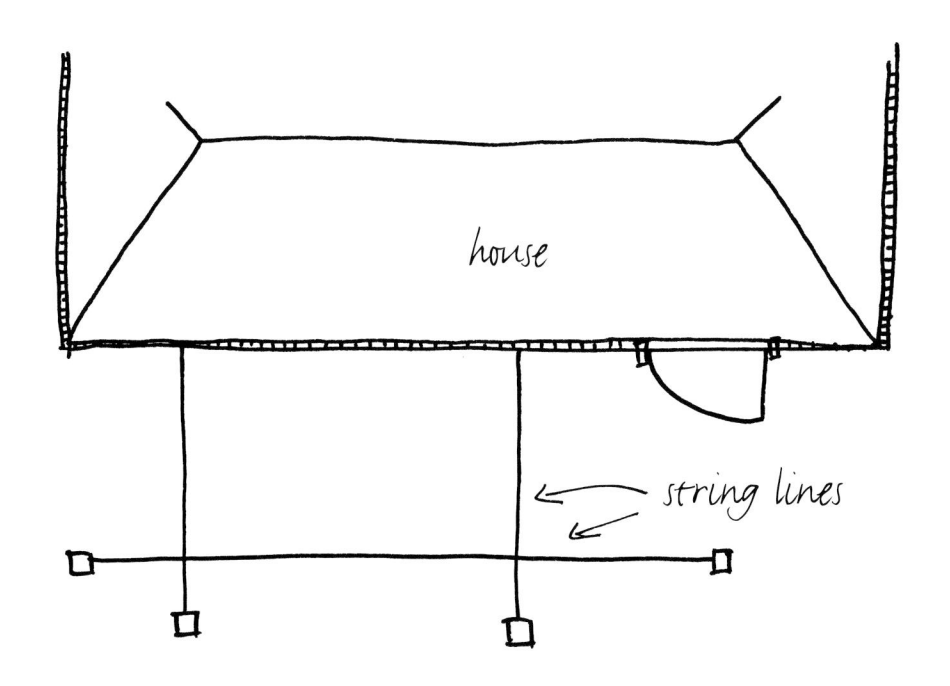

house

string lines

2 Once the string lines are set out, representing the 2400 x 2400 mm (8 x 8 ft) square of the pergola, dig two holes 600 mm (2 ft) deep in the outside corners, ready to drop in two H4 treated pine posts. In this case, the top of the pergola's rafters will be 2400 mm (8 ft), and the posts will stick up another 300 mm (12 in). Depending on how much room you have under the eaves, you may decide to build a lower or higher pergola, which will obviously mean shorter or longer posts. Once you're happy with them, set the posts in the holes with quick-setting concrete (see 'Cementing in a post' on page 18).

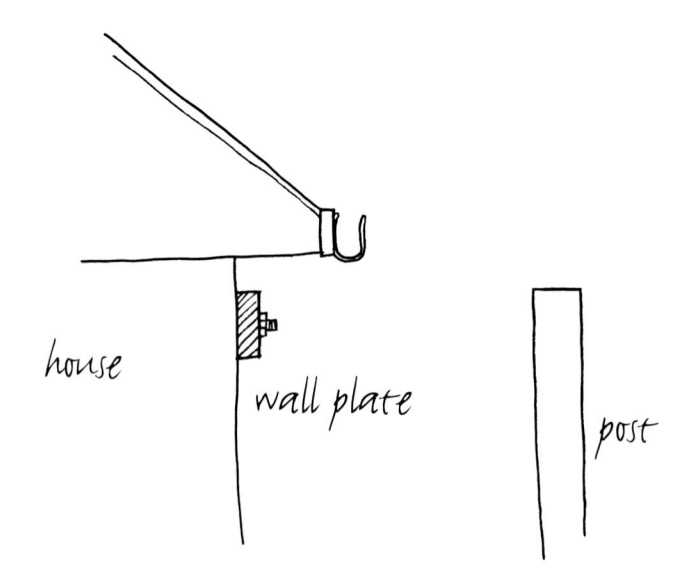

house | *wall plate* | *post*

3 Now it gets tricky. You've got to dynabolt a 150 × 50 (6 × 2 in) timber beam called a wall plate to the house. It has to be level, and at the same height you want the pergola to be. It also has to be exactly the same length as the distance from the outside to the outside of your posts. Once your posts are perfectly level on both sides, and the wall plate is in and level, then you're on fire. The hard part is out of the way.

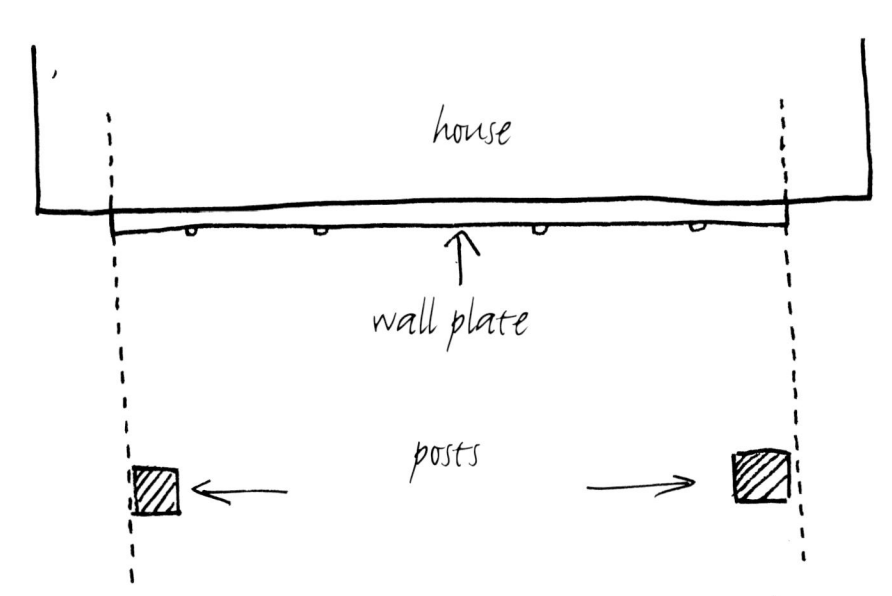

4. Now for the main decorative beam. Cut a piece of 150 x 50 mm (6 x 2 in) H4 to 3000 mm (10 ft). This will give you 300 mm (12 in) overhang on each post.

5. At one end of the beam, measure down 50 mm (2 in) and in 150 mm (6 in). Join these two marks with a pencil line. Do the same at the other end of the beam.

50 mm (2 in)

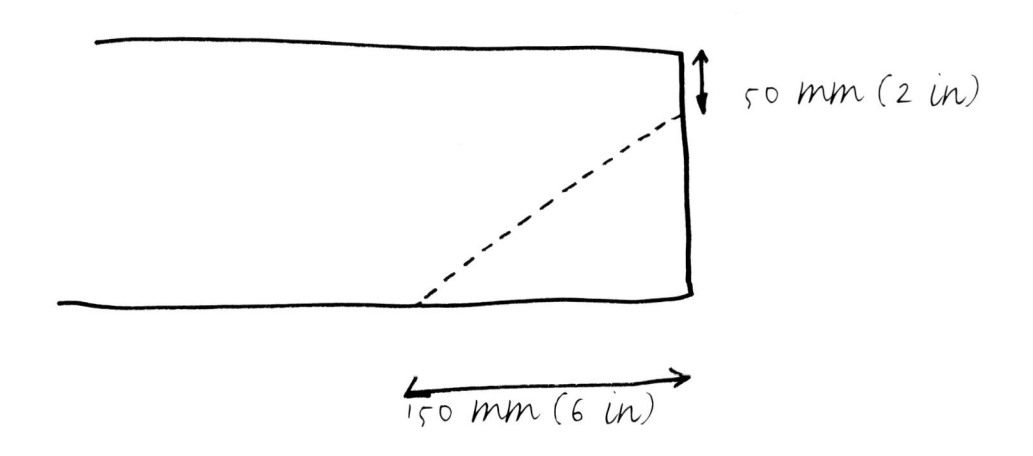

150 mm (6 in)

6 Cut along both these lines, and you have a decorative beam.

7 Using your water level (see page 23), transfer the height of the top of the wall plate to the posts. These marks represent the top of the decorative beam.

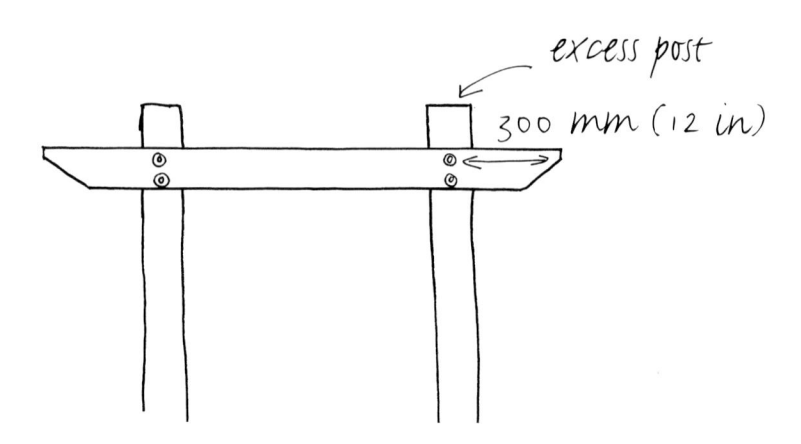

excess post
300 mm (12 in)

8 On the beam, mark 300 mm (12 in) in from both ends. This represents the outside of the posts. Now you have all

YOUR MARKS IN PLACE. GET A MATE TO HELP YOU, AND BOLT THE DECORATIVE BEAM INTO POSITION WITH GALVANIZED BOLTS.

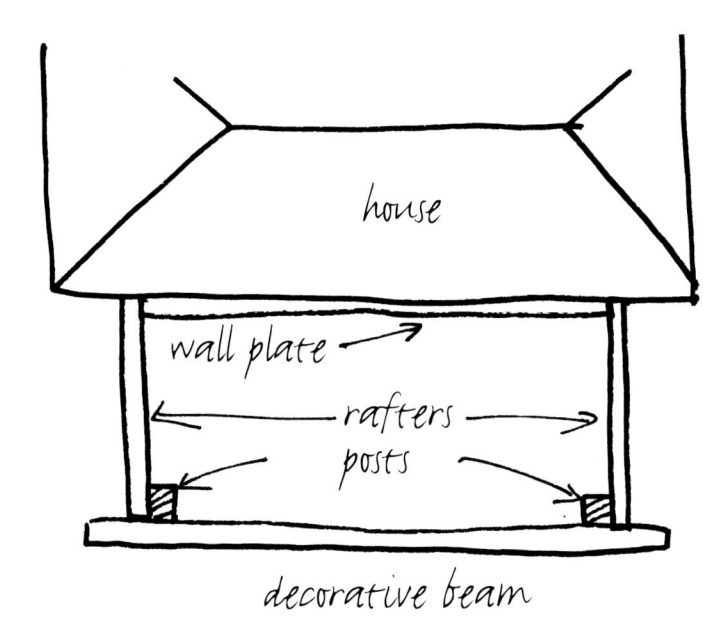

house

wall plate

rafters

posts

decorative beam

9 NOW FOR THE FIVE RAFTERS. THE FIRST TWO GO ON THE OUTSIDE OF EACH POST. FASTEN THESE INTO THE END GRAIN OF THE WALL PLATE. WHEN FASTENING YOUR RAFTERS IT'S A GOOD IDEA TO USE JOIST HANGERS. YOU CAN GET THESE ANYWHERE, AND THEY GIVE YOUR PERGOLA A MUCH STRONGER AND LONGER LIFE. ASK AT THE HARDWARE ABOUT THEM.

10 NOW EVENLY SPACE THREE MORE RAFTERS. THIS SHOULD GIVE YOU A GAP OF ROUGHLY 600 MM (2 FT) BETWEEN EACH

RAFTER. REMEMBER, THE BEAMS, WALL PLATE AND RAFTERS ARE ALL FLUSH AT THE TOP.

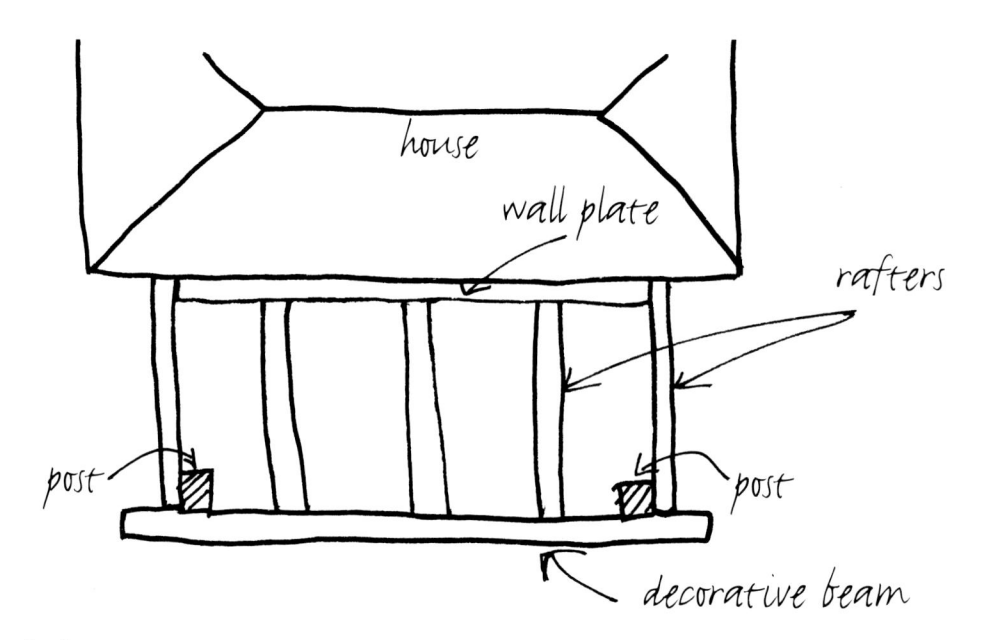

house, wall plate, rafters, post, post, decorative beam

11 TRIM OFF YOUR EXCESS POSTS FLUSH WITH THE TOP OF YOUR BEAM AND HEY PRESTO, YOU HAVE A PERGOLA!

12 NOW THAT YOU'VE FINISHED, AT THE BASE OF EACH POST THROW IN SOME STAR JASMINE OR PASSIONFRUIT. OR YOU COULD CHOOSE THE ONE I LIKE MOST, A GRAPEVINE. I HAD ONE GROWING ONCE AT AN OLD HOUSE OF MINE. SHE LEAFS UP IN SUMMER TO GIVE PROTECTION FROM THE SUN, AND DROPS ALL HER LEAVES IN WINTER TO LET IT IN (OR, AS A GARDENER WOULD SAY, IT'S DECIDUOUS). THE

Grapevine I had gave us about 20 bunches of table grapes a year. They were great.

TIP OF THE DAY: TRADE SECRET

If you want to put a beam up by yourself, nail two offcuts of timber into the posts to represent the bottom of the beam. Sit the beam on top of the offcuts and fasten away. And when they come and say, 'How did you do that on your own?', just say, 'Never you mind, that's called a trade secret.'

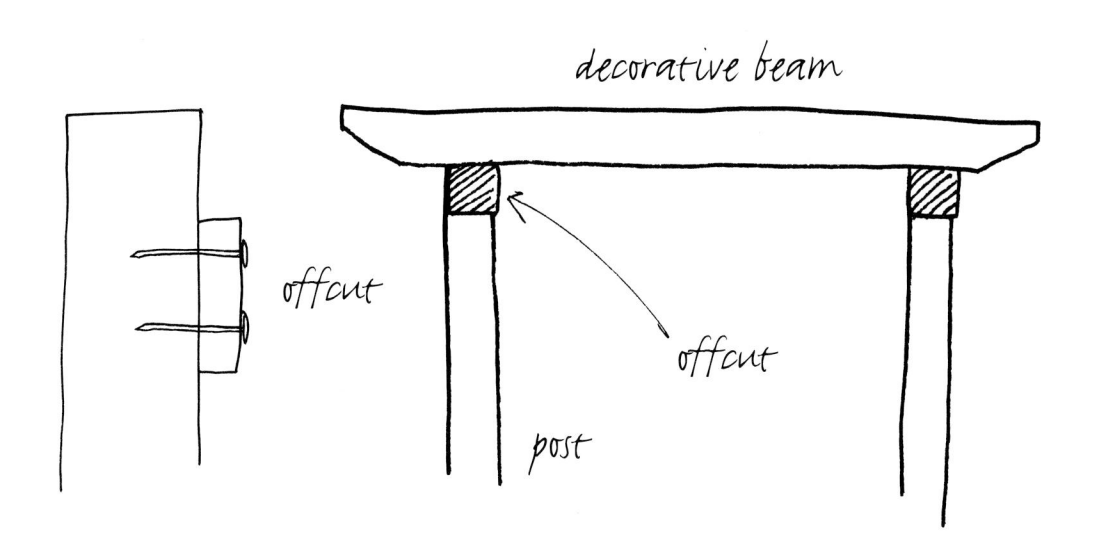

decorative beam

offcut

offcut

post

The sleep clinic

I'm going to admit something very personal. It's private and my wife, the Minister for Elbow Jabs, will be upset with me for giving up something that's in-house.

I snore. Not just any snoring... freight train stuff. I'll suck the curtains off the windows if they're hung too close. I wake up with a sore throat and croaky voice 'cause I've been going all night.

In the old days it was only when I was lying on my back, so the Minister made me a special T-shirt to sleep in. It had a tennis ball sewn to the back of it, so that when I rolled onto my back I was sleeping on a tennis ball.

When she first gave this T-shirt to me, I thought, 'She's done some sort of S&M sewing course at TAFE.' I said, 'I refuse to wear this, no, no. No way.'

Anyway, after the first couple of nights, it wasn't too bad. I got used to not sleeping on my back, and the lack of left elbow jabs was a pleasant change to my night's sleep pattern.

But the wheels fell off the wagon when I began snoring on my side: the tennis ball torture wasn't working. Something had to be done.

An appointment was made, and off I went to the sleep doctor. True to his name, during our interview I started to fall asleep. I thought, 'This sleep doctor is really good.' I half expected him to be in his pyjamas.

First question: 'Do you have a drink?'

'Four or five,' I said.

'Four or five a week...that's not too bad.'

I just looked at him and said, 'Yeah.' I didn't have the heart to tell him it was every day.

'Do you smoke?' In those days I did. I don't now. (Filthy stinking habit, but when you're a smoker there's nothing finer than a smoke. If you've never smoked it's hard to explain.)

'Yes, I do smoke. Nearly a packet a day.'

'OK, first up, if you want to stop snoring, you'll have to give up the drink and smokes.'

Silence for a while, then I just said, 'No. What are the other options?' I mean, this bloke was kidding, wasn't he? Stinking hot summer's day, putting in a good 8–10 hours on the tools, and you're not allowed to have a beer! Yeah, sure mate.

'The other option is you go to the sleep clinic, we check out what you do during the night, and try and come up with a solution.'

'Now you're talking, Doc. See, it wasn't hard to think of something more sensible, was it?'

We made an appointment for 7 pm the next Friday night. I got a bit of a rundown as to what to do, what to wear and what to bring. The most important thing was not to do anything different in the arvo, or leading up to sleep time.

The Friday arrived. I did my normal day's yakka, popped into the pub for a drink with the boys, then home about 6 pm for a quick bite to eat and a beer. The Boss had a bag packed for me, so I went and got the six-pack Esky, whacked in the cans and ice, and I was ready to go. She said, 'What are you doing?'

'Well, I would've had some cans in front of the telly tonight, so I'm keeping the routine going and taking the cans with me. Everything has to be normal.'

At the sleep clinic, I was set up in my room, and then a nurse came in and started attaching wires to me and marking my head, arms and legs with a felt-tip pen. I had 10 wires attached to my face and head, and the same on my chest, arms and legs. I looked like a science experiment. These wires were connected to bigger wires, which went into a small box with a shoulder strap so I could carry it around with me. By 7.30 I was all hooked up.

I flicked on the telly to find Friday night footy – you beauty! Got my little Esky out, wound up the back of the bed and cracked a can. I've always thought those hospital beds would make great lounge chairs. Forget the recliner. Go one step further with the hospital bed: you've got about five positions, those little tables that wheel in and, of course, a buzzer for when you need something, like another beer. Best case scenario: if you could talk the Missus into wearing a short nurse's uniform, you've got yourself the perfect piece of furniture.

By half-time in the footy, I really felt like a ciggie. So I popped out to the nurse to find out how to get to the verandah or outdoor area. There was none. I had to go downstairs to the front of the hospital.

The front was like a city department store on Boxing Day, people everywhere. I only had on a pair of shorts, there were wires and felt-tip pen marks all over me. Basically, out the front of the hospital, I would've looked terminal, with about a week to live. So back to the room to tough it out without the smokes.

The beers were going down faster to combat the fact I wasn't smoking. By can six (along with the few I'd had at the pub and at home), I was in the lift and on the way down. I didn't give a rat's what I looked like. I was having a smoke.

I walked through the foyer, wires everywhere, small box over the shoulder, can of beer in hand. I got out the front and sat down with my ciggy and beer.

Women were keeping children away from me in case I was contagious. Strangers were saying 'hello', nodding and smiling, probably to try and cheer me up. Other people just gave me a pitying look, thinking, 'Look at that poor bastard having his last beer and ciggy.'

Needless to say, I slept like a log and snored the house down. The wires didn't bother me. When I've had a few beers, I can sleep on a clothesline.

In the end, the sleep clinic didn't do much to fix my snoring. But I did end up quitting the smokes.

RETAINING WALLS

If you have a bit of a sloping section in your backyard — or you want to level up an area so you can put in a barby area or paving — then a retaining wall is the go.

I recommend you don't go any higher than 400 mm (16 in). Anything substantially higher than that — get in the professionals.

The easiest materials for a beginner to use are H4 treated pine sleepers, 200 x 100 mm (8 x 4 in) — on edge.

1 First, dig out and level a trench where the retaining wall is to go. Make sure the retaining wall is square or parallel with whatever is existing in the backyard.

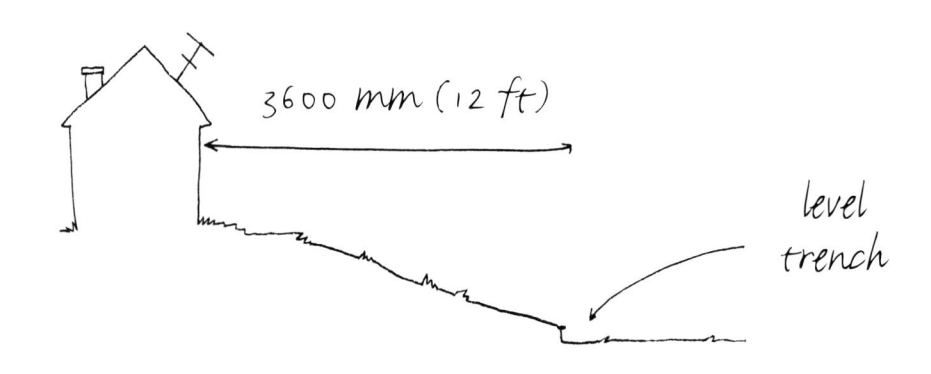

3600 mm (12 ft)

level trench

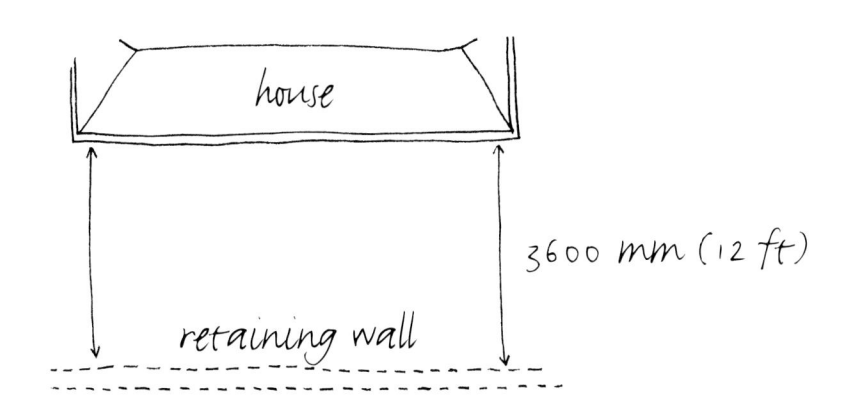

2 Lay the bottom course of sleepers on edge in your trench, making sure as you go that each one is level. The sleeper can be adjusted by packing river sand underneath, then lifting until it is level.

3 Once the first course is laid down and level, it's time for the posts to go in. Use the same material as the wall for the posts — H4 treated pine, 200 mm x 100 mm (8 x 4 in).

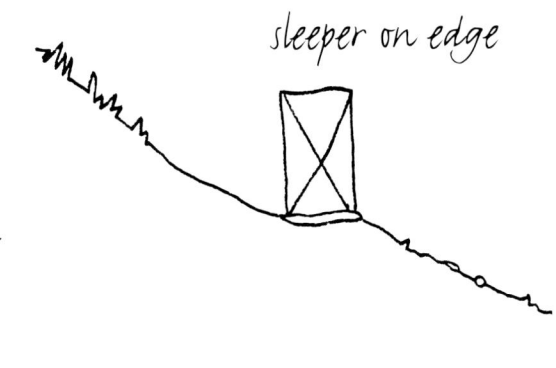

4 Dig a hole wide enough for the sleeper with room to adjust it. Go down at least 600 mm (2 ft). But if the digging is easy (like if there's a lot of sand for instance), dig to at least 900 mm (3 ft).

5 Sit the post in the hole — and check that it will sit level with the retaining sleeper. The hole may need to be adjusted so as to level up the post. The holes should be dug on the joins of the retaining sleepers first. You will be using 2400 mm (8 ft) long sleepers so dig your posts in at 2400 mm (8 ft) centres.

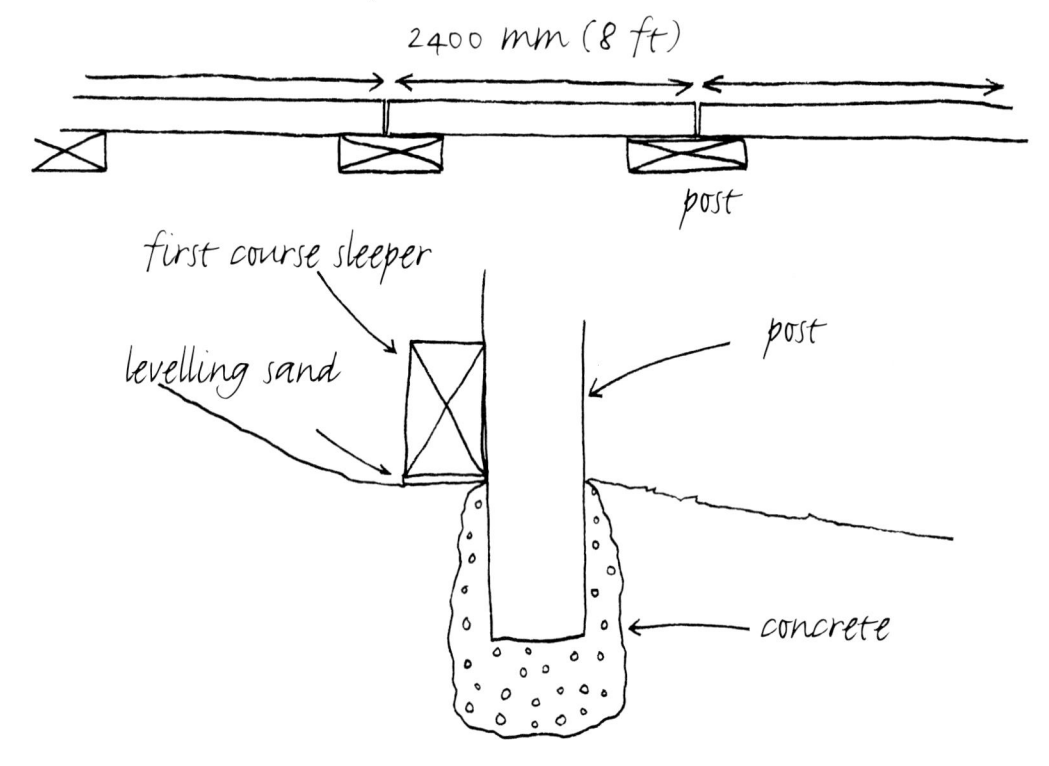

2400 mm (8 ft)

post

first course sleeper
levelling sand
post
concrete

6 Backfill the post hole with quick-setting concrete (see page 18).

56

7 Ensure the posts finish 200 mm (8 in) higher than the top of the first course — this allows for the second course.

8 Now the second course can be put on top of the first. This is a matter of laying it straight on. Make sure the joins fall behind the posts. Whack in a galvanized 75 mm (3 in) nail just to hold things in place.

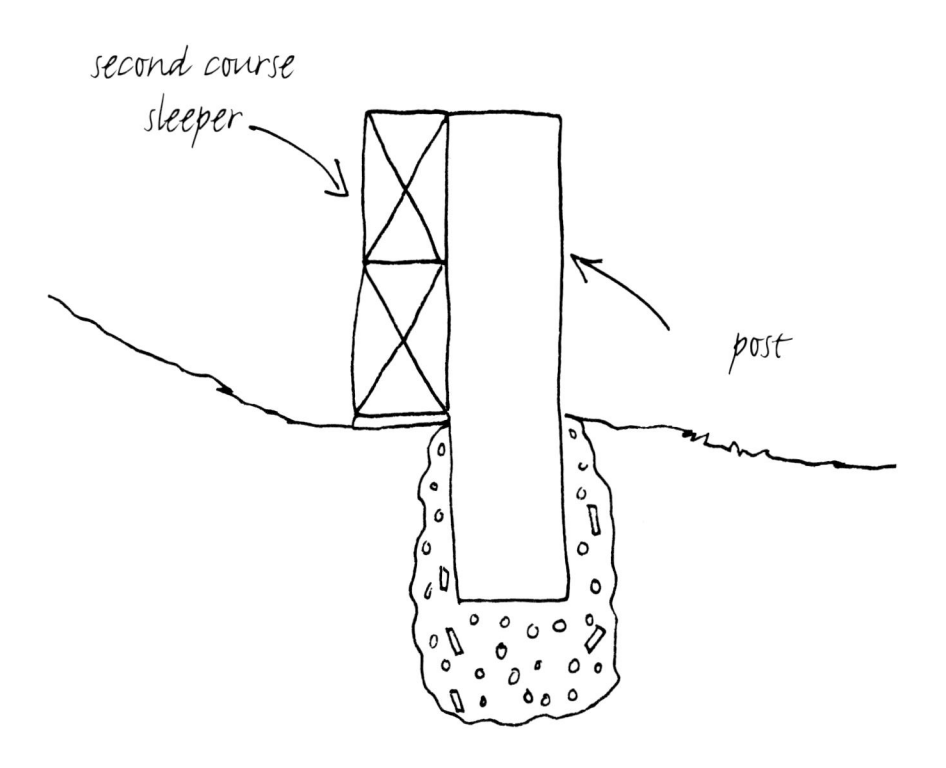

second course sleeper

post

9 Now that the two courses and the joint posts are in, you will need to whack in a few more posts for strength. I usually do them at 1 m (3 ft 4 in) apart, but you will need to even them out so it looks good. Just halve the distance and put a post in between the joint posts so then you have a post centred every 1200 mm (4 ft).

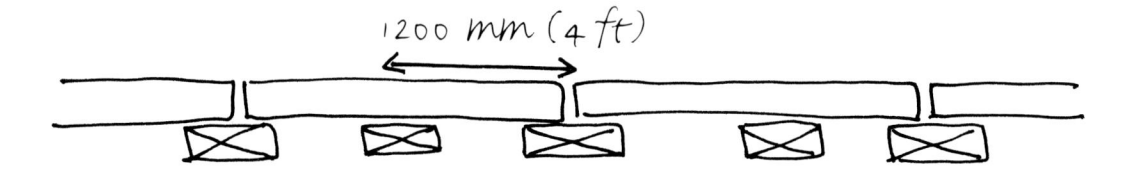

AGRICULTURE DRAIN

The world's best retaining walls have come unstuck because there were no agriculture drains behind them. There is a theory which I am sure has never been tested that you could build a retaining wall out of cardboard if you had adequate drainage — it is nearly always the water below ground which buggers up retaining walls. Of course every now and then it is shoddy workmanship and the wrong materials that are to blame. But generally, when building retaining walls, the agriculture drain is essential.

You have got the retaining wall built, it looks magnificent, the whole family is standing around once again patting you on the back calling you a legend — life's good. Then they say, 'Let's help you backfill it,' and you say 'No! We need an agriculture drain, and here is how you do it.'

Well, here is how you do it.

1 Get yourself a length of agriculture line to match the length of the retaining wall and add about 1 m (3 ft 4 in) to it.

2 You will need to get some blue metal — the black stones you see in concrete. If your retaining wall is quite long you can get them delivered by tipper in bulk — or if your wall is only a couple of metres (yards) long you can buy it in bags.

3 Lay about 50 mm (2 in) of stones on the bottom up against the bottom sleeper — just a thin bed to stop soil getting into the agriculture line.

4 Lay the agriculture line on the blue metal up against the sleeper. The ends poke out past either end of the retaining wall to the lower level.

5 Cut a notch out in each of the two bottom corners of the wall and feed the agriculture line out. It is preferable to hook these up to storm water but run-off into the garden will do.

agriculture line on bed of blue metal

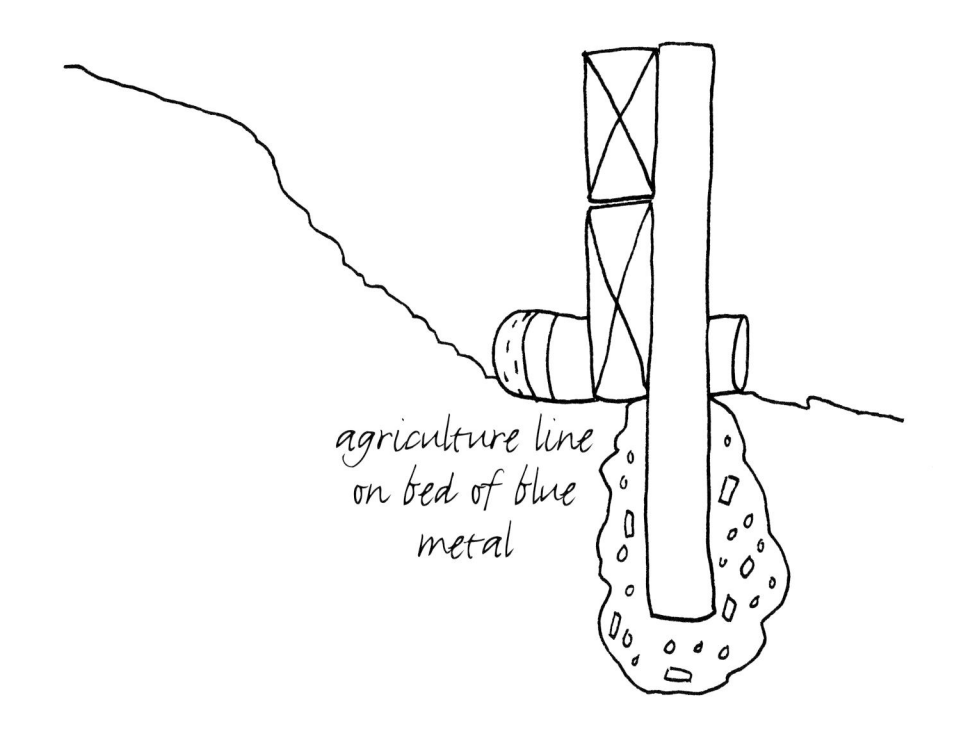

6 Now the agriculture line is in place and either one or both ends are poking out through the bottom of the wall to the lower level, bury the agriculture line in blue metal, covering it by at least 100 mm (4 in).

7 There is a product called Geo-textile fabric. You can buy it off the roll. Lay this down on top of the blue metal, covering it all — this will stop your backfill soil clogging up the blue metal and agriculture line. You can buy a geo-textile sock to put over the agriculture line, but in my opinion they get clogged up and prevent the agriculture line from doing its job 100 per cent.

8 Now that the agriculture line, the blue metal and the fabric are in — just backfill with a nice turf underlay if you're laying turf or just good clean fill for anything else. The retaining wall will give you a level area, which in my book walks all over a sloping one.

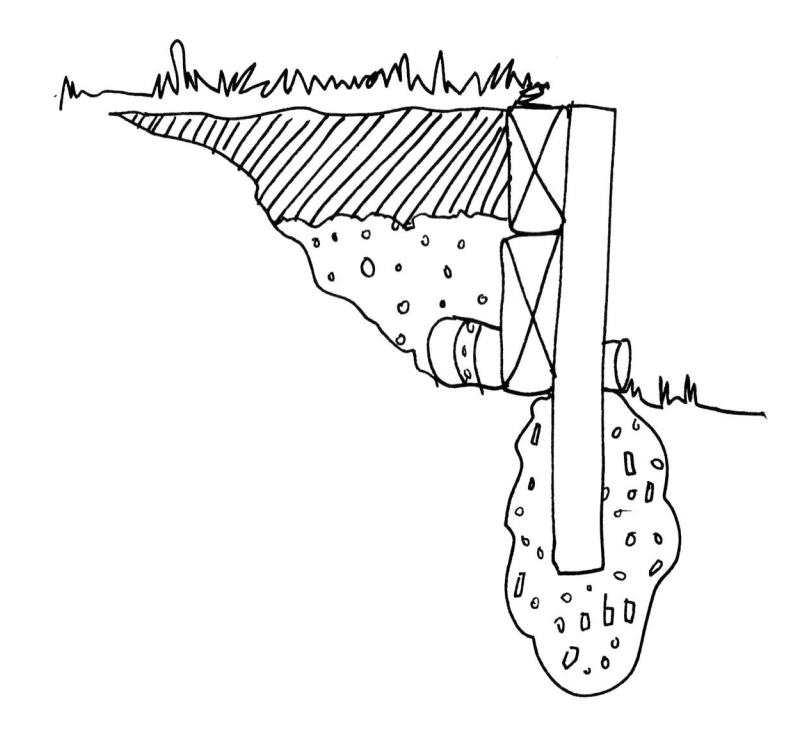

TREEHOUSE

A treehouse is the ultimate hide-out, the world's best clubhouse. I had heaps of treehouses as a kid. The first lot I built, I was the labourer and my brother was foreman. When he moved onto bigger and better things, I became site treehouse foreman and built some absolute crackers, mostly using old fence palings and scrap timber and branches.

To me, they were the Taj Mahal, St Paul's Cathedral and Parliament House all rolled into one (and the construction was much better than under the previous foreman). The treehouse building must have been an omen, as my big brother went on to become a builder and I became his apprentice. I think that's life imitating life.

I've built a few treehouses for my kids and they love them. There is no need for a 'Swiss Family Robinson' job — a simple platform among the branches will do the trick. To a six-year-old, that's a palace.

Platform Treehouse

A really easy treehouse to build is a small platform beside a tree. Make the platform high enough so that it is among the branches. Have a rail on one side, and sling a rope around one end of the rail so the kids can use it for a quick escape. Over the other end of the rail, sling a second rope on a pulley with a bucket on one end. The kids can use the bucket to haul stuff onto the platform — unreal! So, here's how to build a treehouse without using the tree.

MATERIALS

Two 90 x 90 x 2100 mm (3½ x 3½ in x 7 ft) posts, H4 treated pine

Two 90 x 90 x 2700 mm (3½ x 3½ in x 9 ft) posts, H4 treated pine

Two 150 x 50 x 2030 mm (6 x 2 in x 6 ft 9 in) beams, H4 treated pine

One 150 x 50 x 3230 mm (6 x 2 in x 10 ft 9 in) beam, H4 treated pine

Five 100 x 50 x 1940 mm (4 x 2 x 8 ft 5½ in) joists, H4 treated pine

One 1850 x 1940 mm (8 ft 2 in x 8 ft 5½ in) floor platform, cut from two sheets of 18 mm (¾ in) weathershield ply

About 3500 mm (12 ft) of 18 mm (¾ in) rope

About 3500 mm (12 ft) of 9 mm (⅜ in) rope

Pulley to suit 9 mm (⅜ in) rope

Bucket

1 Start by setting out a square on the ground representing the size of the platform. Using string lines and pegs, set out your square or rectangle (refer to 'Do your homework' on page 16 if you need to brush up on your set-out skills).

2 Dig the post holes 600 mm (2 ft) deep, and then use quick-setting concrete to secure your 90 x 90 mm (3½ x 3½ in) posts into the ground. Remember to use H4 treated pine. Two posts should be 1500 mm (5 ft) out of the ground, and the other two 2100 mm (7 ft) out of the ground. Depending on how big your tree or backyard is, you can modify the dimensions if you like. Just make sure the distance between the short and long posts is no more than 1800 mm (6 ft). There should be 1850 mm (6 ft 2 in) between the posts in the other direction.

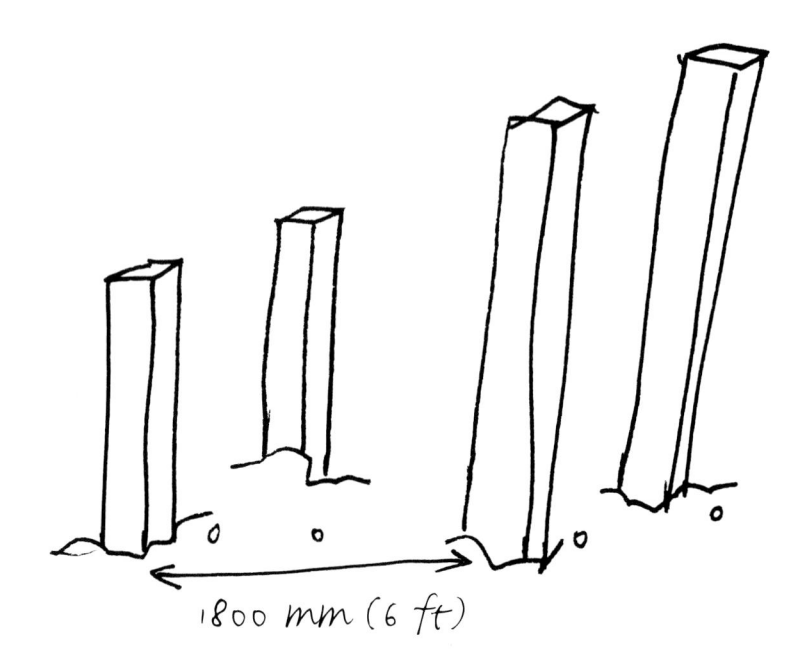

1800 mm (6 ft)

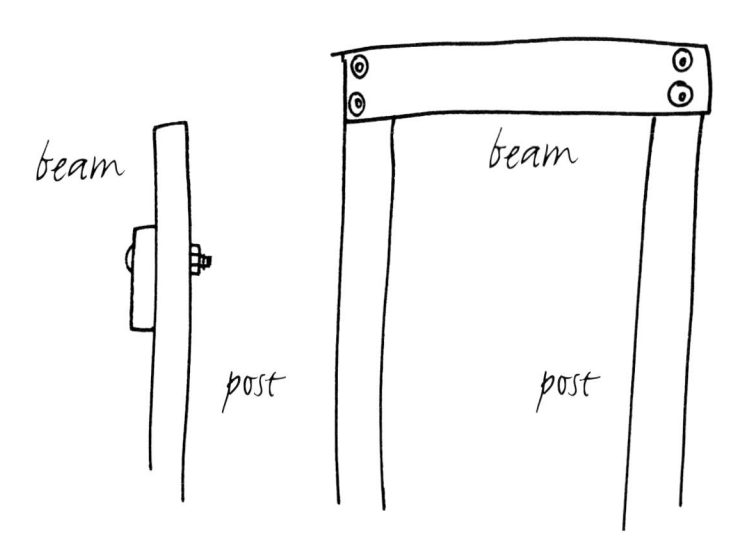

3 Bolt your two 150 x 50 mm (6 x 2 in) beams to the two sets of posts, making sure they're level. Choose a height that's safe for your kids, depending on how big they are. The two beams can only be as high as the top of the shorter posts.

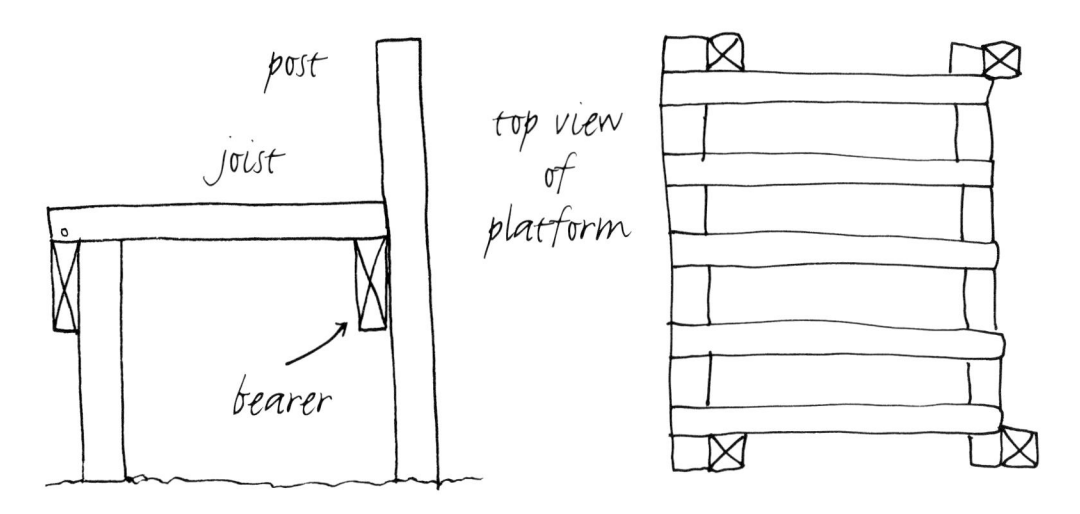

4 Now lay your joists across the beams. These are spaced at 450 mm (18 in)

67

centres. That means the distance from the centre of one joist to the centre of the next one is 450 mm (18 in). In this situation, these 100 x 50 mm (4 x 2 in) joists should not span more than 1800 mm (6 ft); that's why we kept the distance between the short and long posts at 1800 mm (6 ft).

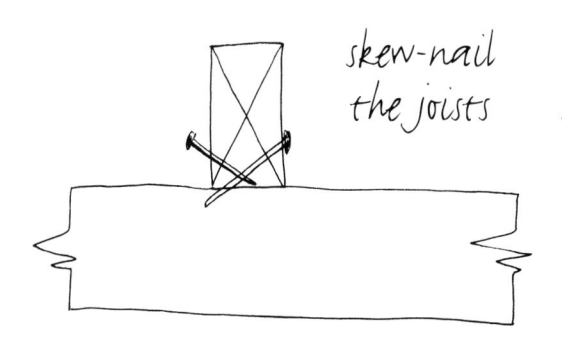

skew-nail the joists

5 Next, skew-nail the joists. Use 75 mm (3 in) galvanized bullet-head nails. This takes a little practice, but before long you'll be an expert.

6 Bolt your third 150 x 50 mm (6 x 2 in) beam at the top of your taller posts. It will have a 600 mm (2 ft) overhang on either side.

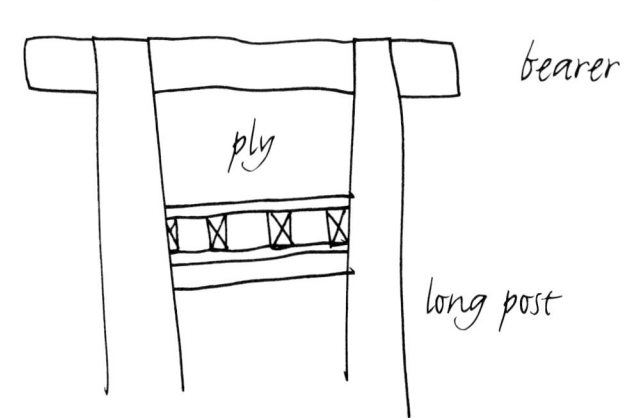

bearer

ply

long post

7 Cut, glue and nail 18 mm (3/4 in) weathershield ply to suit your floor space, and you're nearly there.

8 Cut a small notch on the top side of one overhang, about 200 mm (8 in) in from the end, and tie the 18 mm (3/4 in) escape rope around this. Your hardware might not have rope this thick, but a rope supplier or ship's chandler will. The thicker rope is much better for kids to clamber around on. And the notch will stop the rope from slipping — perfect.

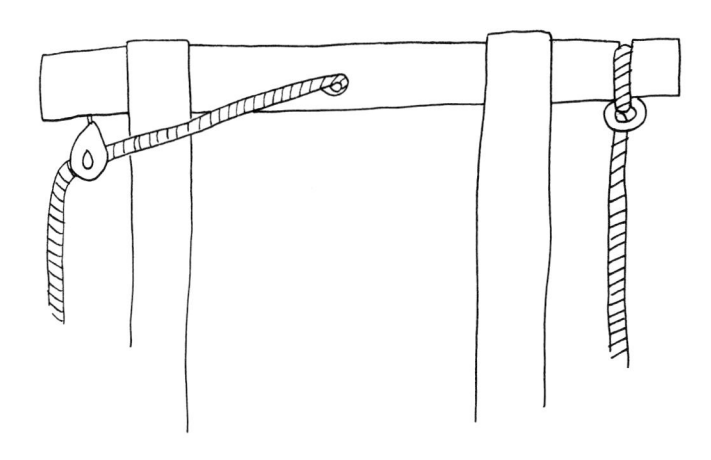

200 mm (8 in)

9 Grab a pulley from the hardware. Screw it to the underside of the other

69

overhang, 200 mm (8 in) from the end of the post, and feed a 9 mm (3/8 in) rope through. Attach the rope to a bucket on the ground and whack a plastic-coated hook into the top beam. Loop the rope onto it so that it's always at hand.

Sandpit

Knock together a sandpit under the treehouse. It adds a whole new dimension, and it's a one-step wonder — too easy!

Materials

Two 150 x 50 x 1980 mm (6 x 2 in x 6 ft 7 in) lengths H4 treated pine
Two 150 x 50 x 2130 mm (6 x 2 in x 7 ft 1 in) lengths H4 treated pine
Plastic sheeting
Clean pit sand

To the outside of your posts, bolt the four lengths of 150 x 50 mm (6 x 2 in) at ground level. The shorter bits go in the same direction as the joists, then the longer bits overlap them. Lay in some plastic and fill the area with sand — instant sandpit!

The True Treehouse

Now a platform built among the branches of a tree is the ultimate hide-out. A 7-year-old will sit silently camouflaged in the canopy for hours.

Materials

The first and most obvious thing you need is — a tree! It needs to be of a reasonable size, with three fairly thick branches, but four is even better.

As far as the other materials go, well, building a treehouse is 100 per cent improvization. You've got to make it up as you go along, and that's what I love about it. However, you can be pretty sure you'll need some 150 x 50 mm (6 x 2 in) beams, a few 100 x 50 mm (4 x 2 in) joists, some 18 mm weathershield ply and a fair bit of 75 x 50 mm (3 x 2 in) for a ladder and handrail. Being outside, all your timbers should be H4. Also, you will need some perforated hoop iron (a metal strip with holes along it).

1 You've got to find a way to get a beam to span from one trunk to another somehow, any way you can, to create a triangle in the case of three branches or a square if there are four. Once you've achieved this, you're home and hosed.

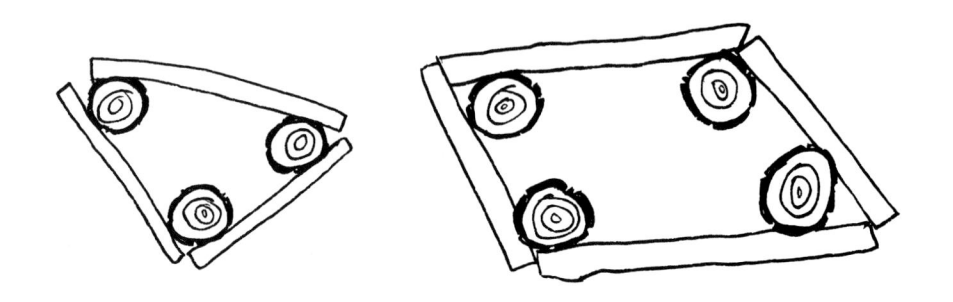

2 Buy a metal strip with holes all through it. It's called perforated hoop

IRON, AND IT'S A GREAT WAY TO ADD STRUCTURAL STRENGTH TO YOUR BEAMS. WRAP THE HOOP IRON AROUND THE PERIMETER OF THE MAIN BEARERS AND NAIL IN CLOUTS TO KEEP IT NICE AND TIGHT. SOME 100 MM (4 IN) NAILS THROUGH THE BEARER STRAIGHT INTO THE TREE TRUNK WILL KEEP ALL THE BEARERS SOLID. THEY WON'T HURT THE TREE, ALTHOUGH WRAPPING WIRE AROUND IT WILL RINGBARK IT, SO AVOID THIS.

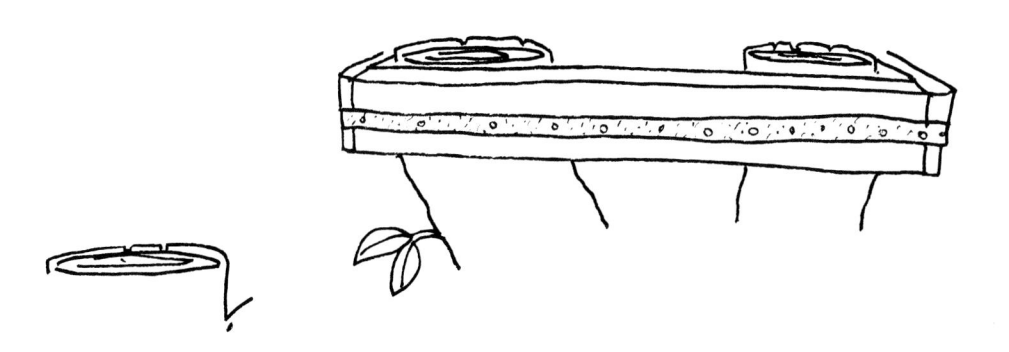

3. NOW THAT THE FRAME AROUND THE PERIMETER IS DONE, WHACK IN SOME JOISTS WHEREVER YOU CAN, THEN NAIL ON SOME PLY OR DECKING BOARDS AS A PLATFORM.

4 Another series of outside beams, about 400–600 mm (16–24 in) up from the platform, will be your handrails. But remember to leave out the handrail on the shortest side of the triangle to make way for the entrance and the ladder.

5 The ladder's really simple. Fasten two lengths of 75 x 50 mm (3 x 2 in) from the ground to the top of the platform, making sure they are parallel and both leaning at the same angle. Cut and bolt 75 x 50 mm (3 x 2 in) steps to the face of the uprights.

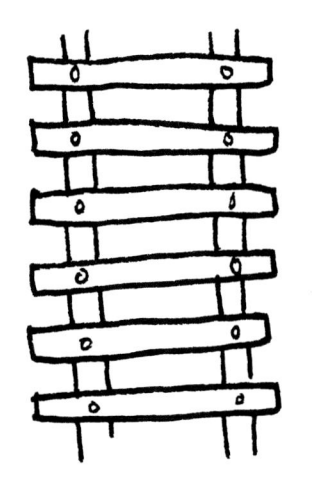

6 The distance between the treads depends on your child. The whole idea of the treehouse is that it should be rough but safe. So having said that, make sure to use a nut and bolt wherever you can.

My young bloke loves his treehouse. I loved mine, and I guarantee your kids will love theirs too.

DOG KENNEL

It's a bit expensive these days to have a mansion of your own. Unless you're man's best friend, that is. This is more than just a dog kennel — it's a little plywood pooch palace. The entire roof lifts off when you need to give it a clean, and the whole thing is perched on three H4 foundations to let a cool breeze blow under it in summer, and to keep moisture from intruding in winter.

Materials
Two sheets 18 mm (3/4 in) weathershield ply
Three 600 x 100 x 50 mm (24 x 4 x 2 in) supports, H4 treated pine
Four 300 x 18 x 90 mm (12 x 3/4 x 3 1/2 in) rafters, H4 treated pine

1 You've got to start from the ground up. Cut out the floor of the kennel, 600 x 900 (2 x 3 ft), but first get your dog to hop on the base for a minute to make sure it'll be the right size. You

Don't want to be building a kennel that's so small your pooch's rump will be hanging out, or so big she'll be curled up in the corner shivering when it gets cold. If your dog is much bigger or smaller than Liz, just tweak the sizes to suit.

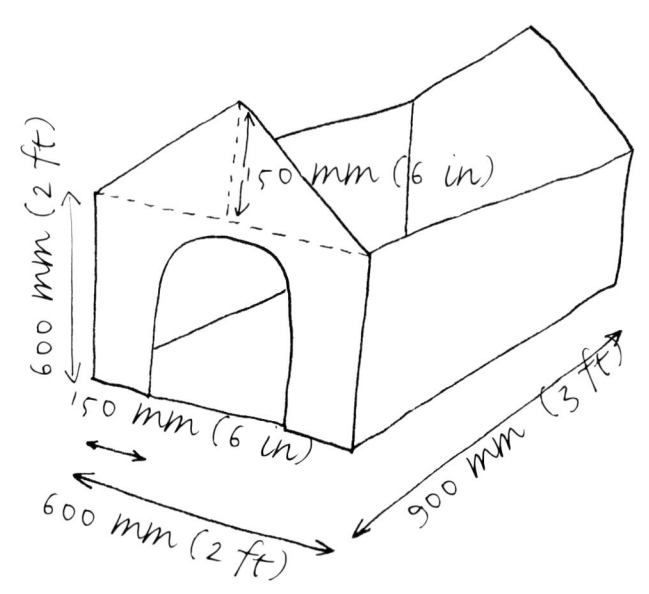

2 Next, cut the front and back walls as in the diagram. The walls are basically 600 mm (2 ft) square with a 150 mm (6 in) tall triangle stuck to the top. The fall of the roof is 22.5 degrees, which sounds pretty scary, until you notice it's just half of 45 degrees. Now that's not so bad, is it? Don't forget to jigsaw out the front door.

3 When you're cutting out the various panels, use what's known as the 'maker's edge' to your advantage. I'm talking about the way all the corners on a sheet of ply are exactly 90 degrees and the edges are perfectly straight.

4 Cut the two side panels, which sit on the base between the front and back walls. They should be 864 mm (2 ft 10½ in) long (so they don't nudge the front and back walls off the base) and 600 mm (2 ft) high. Glue and nail or glue and screw all the bits together — but before you do, draw up the barges on your bit of ply by tracing the angle off the front or back wall. Then when you've got the roof together you can rule off the bottom edge of the barges — remember to leave room for the eaves!

5 Now for the roof. Cut one roof panel 1200 x 450 mm (4 ft x 18 in), the other 1200 x 425 mm (4 ft x 17 in). The second one is a bit shorter so that when you overlap the two, the roof stays nice and symmetrical. You'll need to plane the edges down by a bee's dick so they fit together nicely. To make life easier, just take the angle from the front or back wall using a sliding bevel. It's a nifty little gadget — with this in one hand and a plane in the other, you'll look like a real pro.

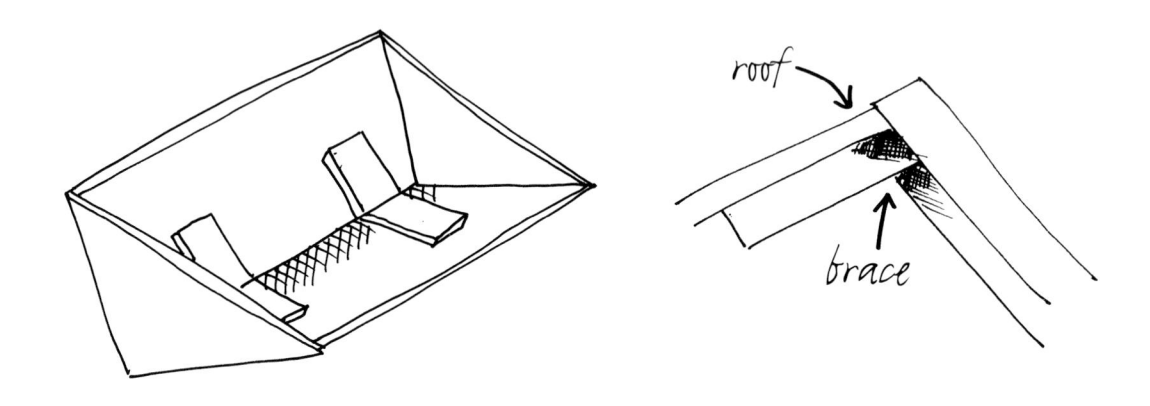

6 You'll need to cut a couple of rafters for each end of the roof. They'll make the panels a bit meatier and easier to nail together, and they'll stop the roof from sliding back and forth. Decking board offcuts are perfect, but anything about 18 x 90 mm ($3/4$ x $3 1/2$ in) will do. Whack each rafter in a mitre box to get a nice 22.5 degree angle at one end.

Glue and nail the rafters to the smaller panel first, remembering to check that the front and back walls fit between the rafters nicely.

7 Then glue and nail the larger panel onto the smaller one — have a squiz at the roof detail to see what fits into where. You can use the kennel structure to help you hold the roof bits in place at the correct angle while you put them together. Then flip the roof on its back again and attach the other two rafters.

8 This is where those lines you drew for the barge come in. Pop the gable end of the roof on the ply and line it up with the angle you drew earlier. Rule across the bottom of your pitch and cut the barges out, then glue and nail them onto the ends of your roof.

9 Screw three bits of H4 treated pine, 600 x 50 x 100 mm (24 x 2 x 4 in) long, to the underside of the kennel. Finish off with a lick of paint.

Dog care

I'm no dog-training expert. I think I'm just lucky 'cause I've got a great dog. Lizzie and I have been together for 13 years. I got her out of the paper when she was about four months old. The ad read: 'Free to good home.' If only those people knew that the dog on the telly was once theirs.

Liz would have to be the smartest dog I've ever had. I know she just lies around all day, but that's because she's used to being on a building site and she doesn't get too excited.

She's never missed a day's work and gets very upset if she's left behind. As I said, I'm no expert on dogs, but I do have a few tips (for what they're worth).

I think to get a good dog you need to be forceful, be in charge. Use a nice loud voice and the same commands all the time, like 'COME 'ERE!' Short sharp words – show them who's boss, and be strict. Once the dog learns, you won't have to be so tough.

And the most important tip is to spend time

with them. The beauty of Liz and I is that we go everywhere together. She sits outside the pub, she sits outside restaurants, she's been in boardrooms, and she's flown all over Australia. She's at her happiest when she's standing beside me.

And that's because, in the early days, I was

always with her. I was able to be on her back all the time, going crook when she did something wrong. The less time you spend with your dog, the more bad habits she gets, and the harder they are to break.

Being hard on the dog early doesn't mean beating the dog up. I gave Liz the odd smack when she was young, but the loud voice scared her more. Training the family dog to be polite makes for a far better life for the pooch.

An obnoxious dog isn't taken anywhere. She's always left at home because she's too much trouble. Liz is a welcome guest at parties and barbies. We take her everywhere, it's a given. So going hard on her for the first two or three years has given her the greatest dog life ever.

I hardly ever bathe Liz because I reckon the more you bathe dogs, the more they stink. They just go out and roll in something to get their dog smell back. Liz doesn't stink — I'm lucky. Other dogs are a different story.

Just about all the dogs I've had are mongrels – a mixture of two good breeds to me seems to produce good dogs. Sometimes dog breeds have been bred a bit finely and now they're having troubles. Just research your breed first, but my tip is the mongrel is the go.

My personal opinion is that the family dog should be a bitch. The girl dogs are much easier to handle and to walk. All a boy dog wants to do is fight and mate, and that's no good for the kids walking him down the local park. So when my mates ask me about dogs, I always say, 'Get a bitch.'

And lastly, give your mutt plenty of pats on the head, and have a special word or two of praise just for them. Mine is 'good girl'. As soon as I say that, Lizzie wags her tail. She knows it's a good tone.

Don't forget: the pat on the head is the international canine symbol for 'thank you'. Dogs love it.

Backyard mainter

Maintenance is the key if you're going to enjoy the yard. My yard was a mess for a while but now it's come up a treat. I keep on top of the maintenance by doing a bit on a Saturday morning, then I've got the rest of the weekend to enjoy it. Remember, a little bit each week will do the job.

THE SHED

The shed can take any form, from the zincalume garden shed to the double garage. It doesn't matter what your shed looks like as long as it's full of stuff you'll probably never use, and it's got a working fridge and at least two items of seating. The seats can be anything from old recliners to a bit of timber sitting up on half a dozen bricks. Just as long as you and at least one mate can sit in there and talk crap with a cold one.

The great thing about the shed is that bullshit doesn't count in the shed. Over a couple of beers, just about any topic or lie is legal.

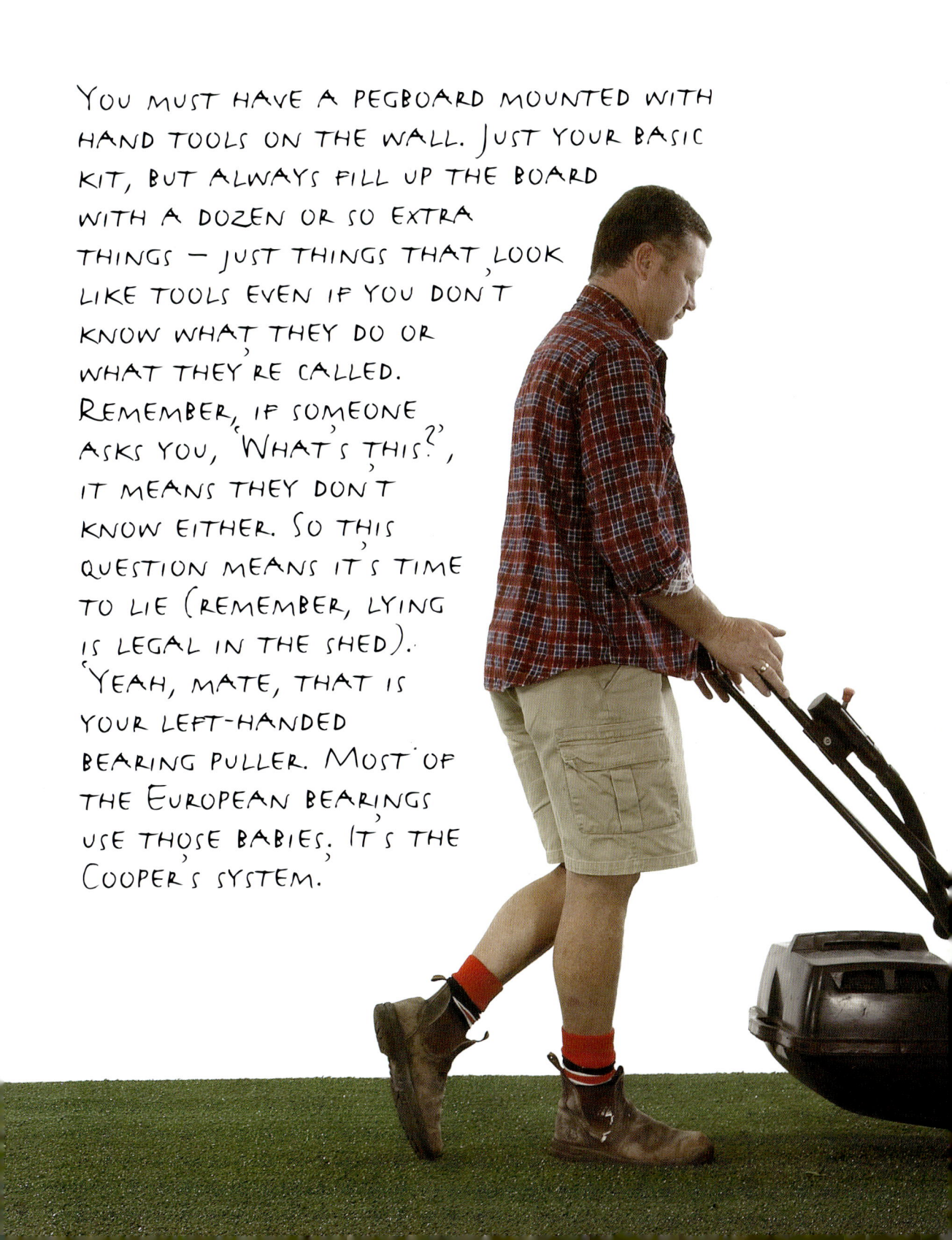

You must have a pegboard mounted with hand tools on the wall. Just your basic kit, but always fill up the board with a dozen or so extra things — just things that look like tools even if you don't know what they do or what they're called. Remember, if someone asks you, 'What's this?', it means they don't know either. So this question means it's time to lie (remember, lying is legal in the shed). 'Yeah, mate, that is your left-handed bearing puller. Most of the European bearings use those babies. It's the Cooper's system.'

Absolute crap, but jeez, that sounded good. Deliver it with a straight face, and it's a flawless pork pie. I know 20 mates who would fall for that straight up, without batting a lid.

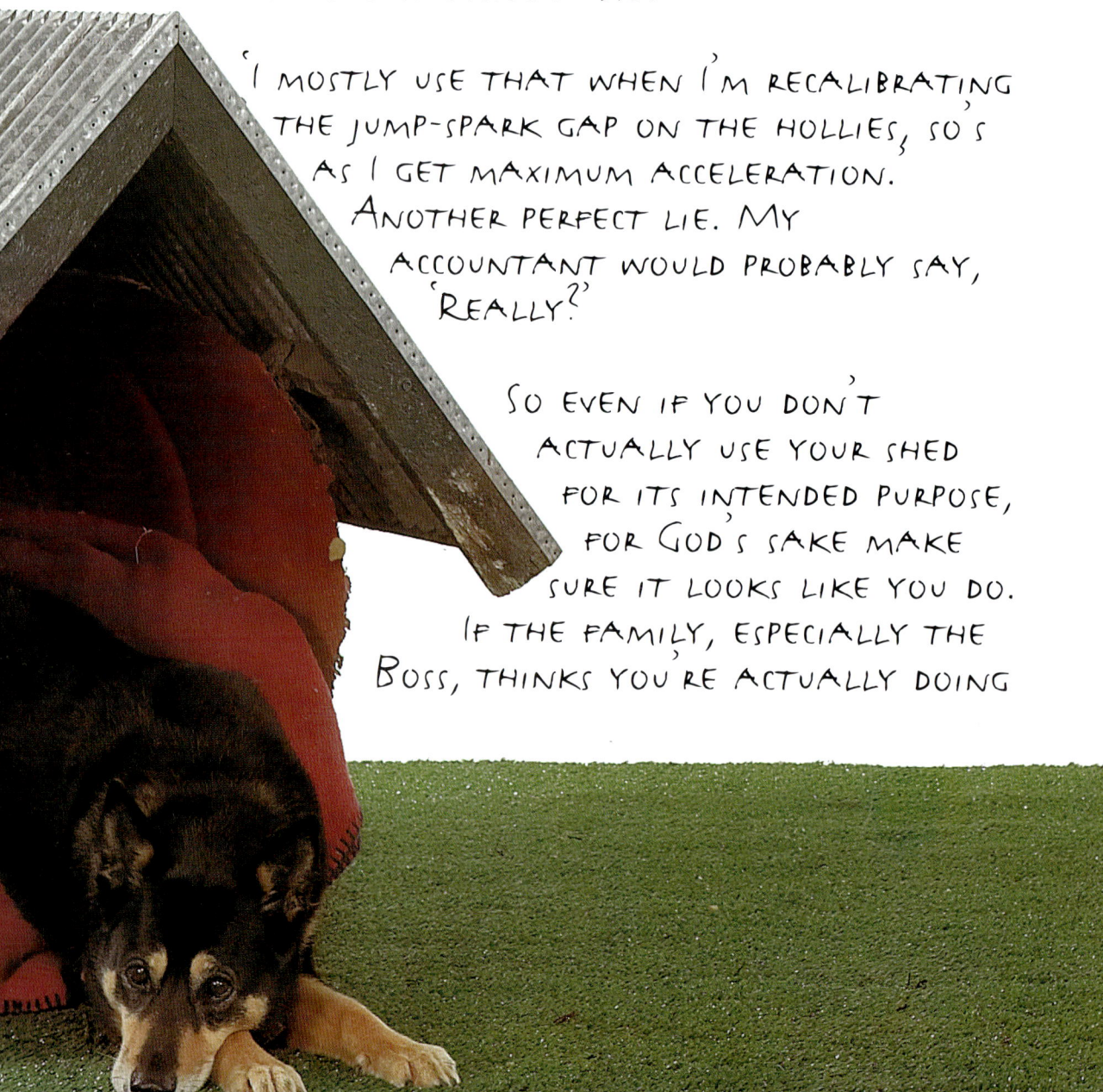

'I mostly use that when I'm recalibrating the jump-spark gap on the Hollies, so's as I get maximum acceleration.' Another perfect lie. My accountant would probably say, 'Really?'

So even if you don't actually use your shed for its intended purpose, for God's sake make sure it looks like you do. If the family, especially the Boss, thinks you're actually doing something in the shed, then there's more chance you'll be allowed out there.

A little colour telly is a cracker, and a radio is the go. An essential item is the Good Sort calendar. It's always handy to know what day it is, I think most blokes would agree...

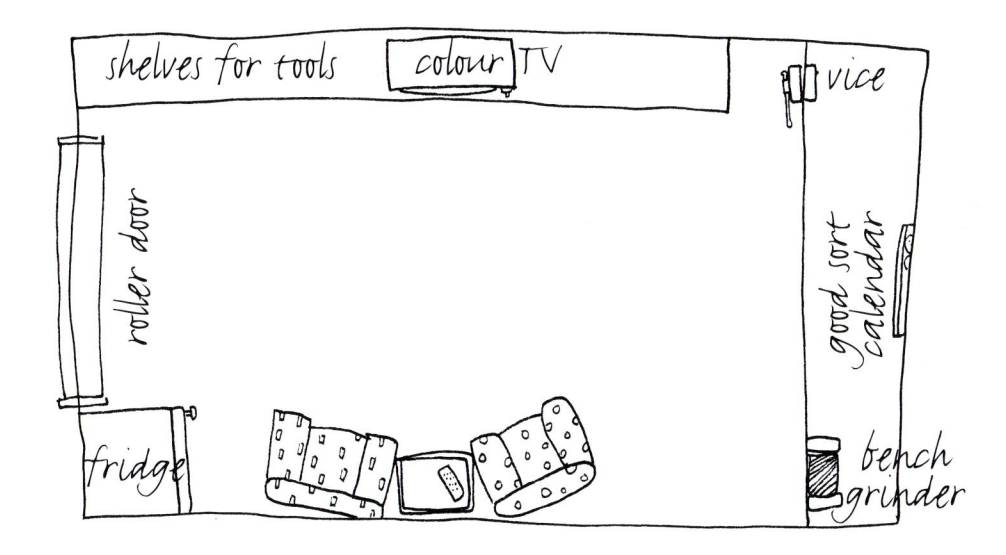

[Diagram of shed layout showing: shelves for tools, colour TV, vice, roller door, good sort calendar, fridge, couches, bench grinder]

Here's a rough layout of my shed.

The boat

Every Thursday I used to race to the newsagency and grab the second-hand bible ('The Trading Post'. I was just like the bloke in 'The Castle' – a 'Trading Post' freak. My mother saw 'The Castle' and said, 'There's a guy in the movie who loves "The Trading Post". He must be based on you.

What a great day when the new bible came out each week. I had a bit of a system: start with boats, straight to the letter 'C', looking for classic timber boats under $500. (Weren't many of them around.) Then to letter 'T' for 'timber boat'. Next, I'd go on to 20 or 50 other categories: second-hand building materials, utes under $500, freebies, miscellaneous, stuff under $20.

One Thursday, something jumped off the page at me:

CLASSIC TIMBER FISHING BOAT
½ CABIN CLINKER HULL REASONABLE CONDITION
$800 ONO.

I'd won the lottery! Fifteen years of searching for the perfect vessel had paid off, and I couldn't believe my luck with the ONO.

Looking back and knowing what I know about boats now, I was an imbecile. I think my wife might have told me that, but that classic line always comes out: 'I know boats.'

So I rang this guy about the boat in the ad. He was surprised to hear from me. And I was surprised that the boat hadn't been snapped up. We arranged to meet to check out the boat. While I was driving there I had images of barbies and bikini-clad women, taking a swim off the bow...

In my mind, this 20-foot, 50-year-old boat had become Greg Norman's latest purchase. I'd turned into Aristotle Onassis, so me and my mates were going to be cruising off the Greek Islands with beer, prawns and birds.

I walked down the hill towards the water and there was a bloke on the shore waiting. I looked at the 100 or so boats moored in front of me. My mind played tricks on me: 'That's it for sure', or 'Maybe that's too big', or 'It could be that one'.

There was one very small, ugly, embarrassing boat. I thought: 'It had better not be that one.' But I knew deep down in my heart that that was it. I just didn't want to admit it, so I kept fantasizing until that actual moment when I met the man on the shore and he said, 'There she is.' My head went down, but only for a split second. The romance of the timber boat, the dog on the bow, the sound of the diesel engine puffing away, a black beanie on, the smell of diesel fumes and teak...

'What a little beauty!' I said. Yep, I talked myself into it — $700 later I was the proud owner of a floating shit box.

Now came the time to move it from its present mooring to the one I'd teed up with a mate of mine. I had to move it from Balmain to Rose Bay. A mate of a mate of a mate had a spare mooring at Rose Bay and he needed to have a boat on it. Years ago you could sell a $1 share of your boat to the guy who owned the mooring and everything would be sweet with the Maritime Services Board. And bloody cheap (about $180 a year at the time — just perfect).

Just had to get it from Balmain, under the Harbour Bridge, through the ferry lanes and up to Rose Bay.

The old diesel started first crack. I was on fire, beanie on, dog on the bow. It was all coming true.

The bloke I bought it off had told me I would need to clean the arse — he said there was a little bit of weed growing on the hull, which would slow it down a fair bit.

I took off, and in the calm water of Birchgrove in Balmain it was going slowly but OK. I was loving it. Cracked open a can to celebrate my first boat, imagining I was the old sea dog. Then, once I was going under the Bridge and coming around the point, things started to go pear-shaped.

The new boat was struggling through the wake of the ferries and there was a strong head wind. We were doing about 1–2 knots, which is not even fast enough to get out of the way of bigger boats.

We were getting tossed around like a cork. I was crossing Circular Quay in the evening peak at 5 pm, doing 2 knots. The dog turned around from the bow of the boat and just looked at me. Shit! Even the dog thought I was a knucklehead.

If I'd had a radio, I would've given it the 'Mayday Mayday' business, but we had to push on. I made it through the Quay and past the Opera House, then, as we worked our way past the Garden Island dockyards, where all the navy ships dock in Sydney, the wind just picked up tenfold. I could see 18-footers capsizing, yachts on massive leans.

Despite the motor going flat strap, we were blown into the naval docks. I had it at full throttle heading out, and I was getting blown in at about 3 knots.

At the time Garden Island had about six US Navy ships in dock. One of them, 'USS Constellation', was the biggest aircraft carrier in the world (if it wasn't, it should've been). Greenpeace had been giving the Yanks some grief as they arrived — you know those flips who run a 10 ft rubber ducky under the bow of a 1000 ft aircraft carrier. Anyway, each to their own.

Now I was being blown straight into the side of the 'Constellation'. The wind was overpowering, the nose of my boat was butting into the side of this massive ship. I had the engine screaming in reverse, and I was standing up in the bow, trying to push off the 'Constellation' and not scratch her paintwork.

Then this Yank leaned over the rail of the ship with a megaphone. He was about 30 storeys up.

He said in a full-on Yank accent: 'Move away from the vessel, sir, or you will be charged under the so-

and-so Act and your boat impounded.

I had the motor screaming, the dog barking, this guy yelling crap at me with the megaphone... I was fairly stressed. I screamed out, 'If I could, I bloody well would, ya knob!'

Eventually I made it out of the Garden Island dockyards and was blown into Rushcutters Bay, where I immediately found a slipway, the first one I was blown into. I had the arse cleaned, plus a few other things done. I worked on it for four days straight and spent about $600. The first of many problems.

But we did have a lot of fun on it for two years, until I sold it, with the motor not working, for $700. I always fooled myself into believing that I'd got my money back.

TIP OF THE DAY: WORK BOOTS

I've always got two pairs of work boots on the go all the time. In the early days I'd bust the arse out of my work boots, then go and buy a new pair and get straight into them. Next thing I knew my feet were on fire and it'd take weeks for them to calm down after breaking my boots in.

Nowadays I have two pairs. Once the old pair is on its way out, the new pair comes in one day a week, or on a Saturday. By the time the arse falls out, the new pair is broken in.

FENCING

There are two types of timber you can use for fencing — hardwood or treated pine. It's up to you, depending on the look you want. The way the fence is constructed remains the same, regardless of what you use.

You'll need morticed posts, rails and palings. The bloke at the timber yard will help you with quantities.

The most important rule in fencing is to be straight and level. That is what will stand out when the job's done. So string lines are very important. Whether you're constructing a new fence, or fixing up a section of an old one, here's what you need to do.

1 Start off with a post in each end of your fence run, then you can use the string lines for the other posts in between. The first post must also be the correct height. So use a paling to make

sure your post out of the ground is the same height as the paling you're using. Whack a post in at each end of your run. If your run has a low point, a third post must go there as well. Let's assume your ground is level. If your ground is all over the place, it's much harder to get your fence level and straight. Just be aware of that if you're gonna have a go.

2 Once the two posts are in, stretch a height string line between the tops of the two posts.

3 Then rig up a side string line along the side of the two posts about 250 mm (10 in) up from ground level. Use the side string line to keep your posts lined up.

4 Now dig the other post holes 2400 mm (8 ft) apart and in line with the string lines. Make sure the post holes slightly encroach past the string line so that the post can be levelled flush with the string line. Here's a bird's eye view of the site.

top string line
morticed posts
post holes
2400 mm
side string line

post   *post holes*   *post*

string line

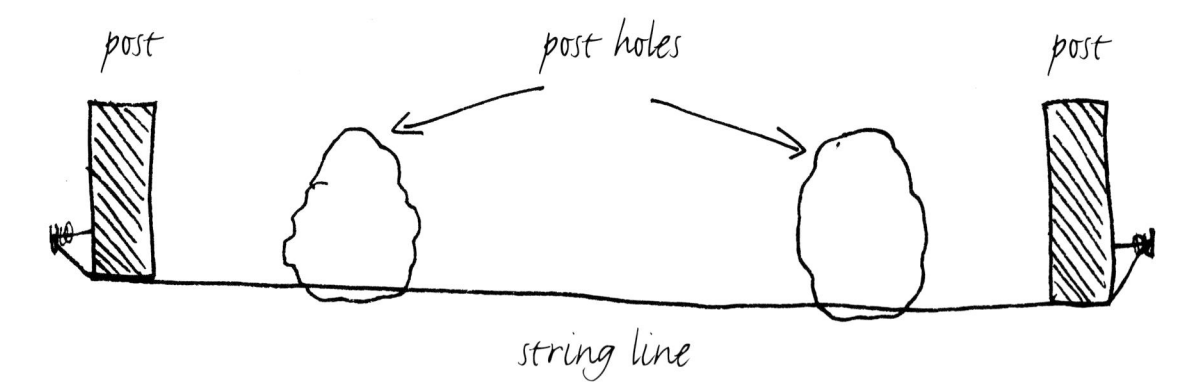

5 SET ALL YOUR POSTS IN CONCRETE, LEVELLING BOTH SIDES AS YOU GO. (SEE 'CEMENTING IN A POST' ON PAGE 18.)

6 THEN PUT IN YOUR RAILS. THE JOINS OF THE RAILS ARE ALWAYS MADE IN THE MORTICE HOLE OF THE POST, AND ARE CUT ON A VERY ACUTE MITRE.

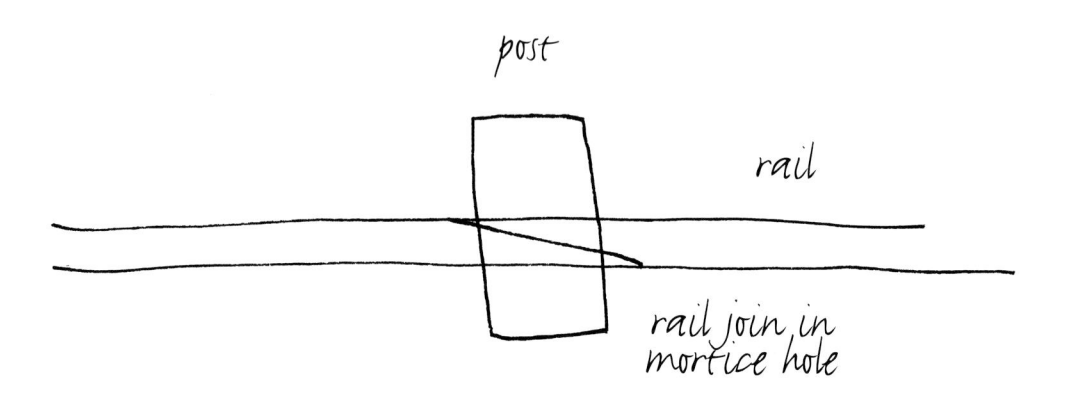

post

rail

rail join in mortice hole

7 Once the posts and rails are in place, you're on fire. It's a good idea to leave your post and rail overnight, 'cause whacking on the palings tends to rattle the posts a fair bit. So let them stiffen up a dash. The best way to attach the palings is with a compressor and a fencing gun, which you can hire for the day. It's a heap easier and you'll be on the cans, getting a pat on the back, a lot quicker. Cut an L-shaped hook from a piece of 9 or 12 mm (3/8 or 1/2 in) ply, and hook it over the top rail to get your spacings between each paling right. Keep your string line in place and work the top of the palings to it.

8 About every fifth board, whack a level on and adjust accordingly. If you're a little bit out, adjust it over the next 5–6 boards, moving just a small amount per board, then keep an eye on it with the level.

There you go — one fence! You legend.

BUILDING A GATE

The gate is always one of the first things to go, 'cause it cops the most activity. You can build a gate with pickets or with a pattern over the top. That's the beauty of it — when you're building it yourself, you can do what you want.

Here's how to build a picket gate. Use this information to replace busted braces or pickets on an existing gate. You can use treated pine or hardwood, but if you're fixing an existing gate, remember to match the materials.

1 First, measure the distance of your opening top and bottom. With old gates these measurements rarely match. If the top of the opening is 805 mm (32$^{3}/_{16}$ in) and the bottom is 812 mm (32$^{1}/_{2}$ in) (see the diagram above right), deduct 10 mm ($^{3}/_{8}$ in) from each distance to allow for 5 mm ($^{3}/_{16}$ in) at both the hinge side and the lock side so that the gate will be able to swing freely.

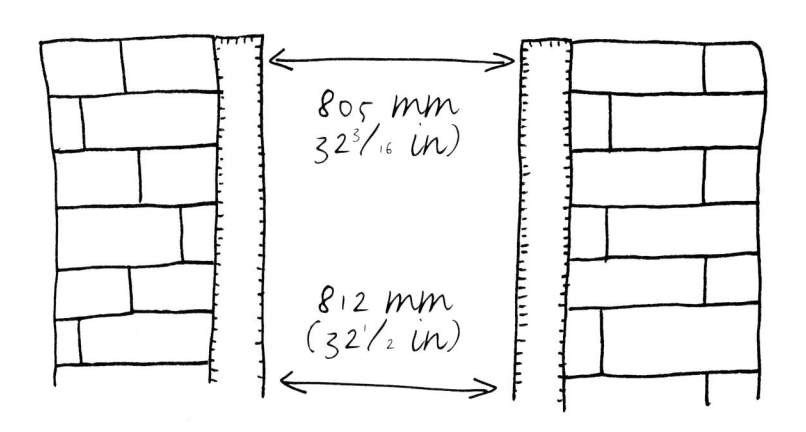

2 Now cut two braces, one at 795 mm (31⁵⁄₁₆ in) and one at 802 mm (32⅛ in). Lay these out on the bench. On each of your two end pickets, 100 mm (4 in) down from the top, and 100 mm (4 in) up from the bottom, lay your braces on your pickets and screw each end in place.

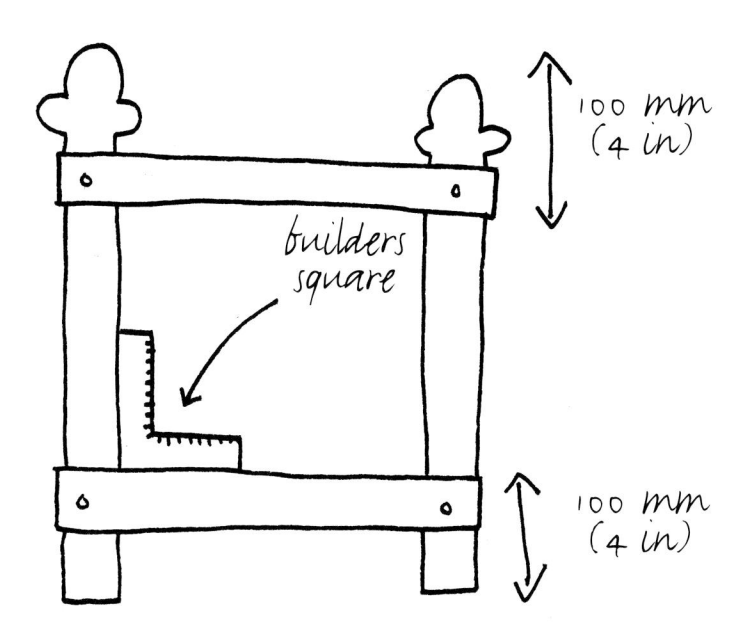

103

3 Make sure the frame is square.

4 Now lay the rest of the pickets, spreading them out evenly with the required gap. Make sure the bottoms of all the pickets are even. If you butt them up against a block of timber you've nailed to the bench, they'll be perfect.

5 Screw all the pickets on, then sit the gate in position to check that it fits. You may need to tweak it a bit. There will be some movement in the squareness of the gate, so make it fit perfectly.

6 Now back to the bench to cut a diagonal brace. This brace must go from the bottom of the hinge side to the top of the lock side. This way you've got a strong triangle of timber between the two hinges and the lock. The brace then helps balance out the forces acting on the gate and overcomes its tendency to sag or twist.

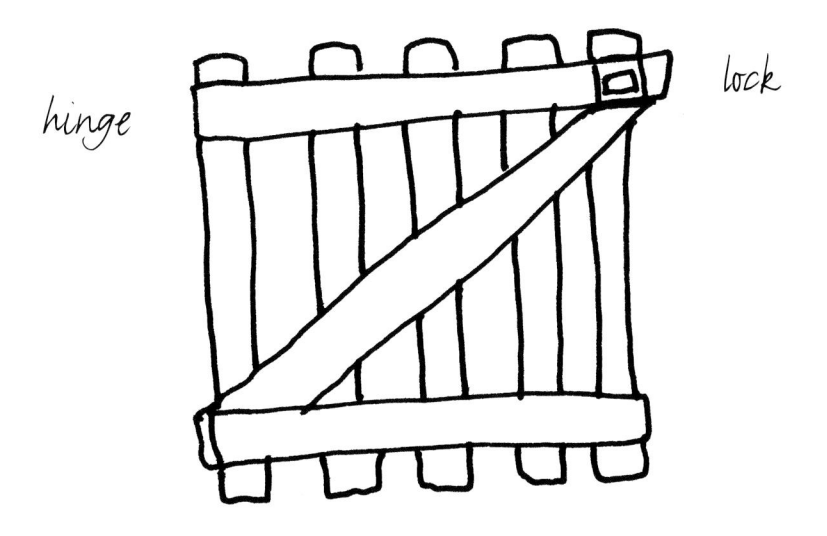

7 Screw each hinge to the gate in line with each cross brace.

8 Screw the lock in the top brace. Then hang the gate back on your post and mark exactly where the lock goes on the post. If you've got a barrel bolt, all you need to do is drill a hole in the place you've marked. If you've got a fancier lock, use the mark on the gate to help you screw the receiving bit in place.

The carpenter and the accountant

One of the great organizations in Australia I believe is the Royal Flying Doctor Service (RFDS). I almost needed them about 15 years ago when I fell through a window and sliced open my wrist fairly badly; the doctor who stitched me up said I had just missed both tendons. We were in a remote part of northern Western Australia, so if I had cut those tendons, I would have been off with the (RFDS) to Perth.

Having worked all over Australia in some pretty out-of-the-way places, I know how necessary the Flying Doctor is. So having said all that, a few mates of mine were going on a car rally called the RFDS Outback Trek to raise money for the service. They needed a support crew of two blokes, with a late model 4WD (that's me) and the crew members had to be either mechanics or doctors or both (that's me). I signed up and had such a great time, it was the first of four treks I went on. Some guys have done about 20 treks, raising millions of dollars. Hats off to 'em.

I needed to find a partner for my vehicle and, being a chippy, I was looking for a doctor or a mechanic. I knew a few doctors, but they were flat out, and I think that maybe spending a week trapped in a ute with an intellectual like myself might have put them off a tad.

I usually just argued with the mechanics I knew. So the only bloke I could come up with was Carmo, who'd been a mate of mine for donkey's years. He was available, and he was willing to spend 10 days in the ute with me as well as share 7000 km (4350 mi) of driving duties. The only problem was, Carmo is an accountant. We'll cross that bridge when we get to it.

The car we were supporting was a 1962 Chevy BelAir, with the car number of 69, so our support vehicle was called X69. We were branded X69 with big stickers on the roof, doors and rear window. Most of the cars carried lollies and small gifts for the kids in the remote towns. The majority of drivers were male, but there were a few girls who were great fun. There were plenty of late nights swagging in the middle of nowhere, drinking and carrying on. It was a real boy's week away, but at the same time it was for a great cause.

Not only was our support car covered in X69 stickers, but it was also plastered with 'Scott Cam Carpenter and Joiner' signs. Basically, the game was up for me. We only had Carmo to work with. But Carmo was notorious for cracking under pressure, and he wasn't a good liar, never had been. We were in trouble.

After a fairly big night in a pub in Condobolin, the town where the Trek started from, Carmo and I turned up at the start line. Blokes and cars everywhere. There were some very professional support crews there as well; some utes were carrying full mechanical workshops in the back. I had my toolbox I used for fixing the cars at home, and I knew a little bit about fixing cars, but I was out of my league. My mechanical skills were limited, but if your door was sticking, now that was a different story.

Once again I was doing what I'm very good at — bullshitting myself, thinking, 'We can get away with this, no one will know.' I looked around to check out some of the support crews and their equipment: it looked like the pit lane at a Formula One race track. 'Anyway, Carmo, just blend in, and like I said, we'll cross that bridge…'

We were flagged off the start line and away we went. No one asked us for credentials. So far so good.

But the wheels literally fell off the wagon when the Chevy threw a bearing about 100 km (60 mi) into the Trek. The idea was that, if someone had a problem, everyone in the general area would stop and look to see if they could help.

So we came round the bend to find about five cars' worth of blokes all looking at the Chevy, which was already being jacked up. They all looked up at us, and you could see the relief on their faces as they collectively said, 'Good, here's their support crew.' As we went past them to find a spot to park, I could see them thinking, 'Oh, they must have borrowed that 4WD off a carpenter.'

There's not too much more to explain about this part of the story. They found out I was a chippy and Carmo an accountant. The boys were all terrific blokes, and had a good laugh. In the end, we came in handy later on in the trek, pulling our weight by towing about 30 cars through a bog. At the end of the day, no one minded as long as you had a go.

As we drove throughout the day, we came to realize that this was a casual affair. Cars would stop in groups of five and six for beers throughout the day, but one person was always kept dry as the driver for that day. After about 300 km (190 mi), I pulled over and said to Carmo, 'Your turn to drive, I need a break.'

Carmo said to me, 'I can't drive a manual.'

'Course you can.'

'No, I can't.'

'What do ya bloody well mean, you can't drive a manual? Everyone can!'

In the 35 years I'd known Carmo, and I suppose, more importantly, in the 17 years we'd been driving, I had never asked him if he could drive a manual. I realized I'd never seen him drive a manual.

It's like asking a bloke if he breathes in and out, or does he enjoy sex, or does he follow the Roosters. (Carmo does.) It would be the last straw if these blokes, nice or not, found out the accountant couldn't drive.

So anyway, we went off by ourselves and Carmo jumped into the driver's seat. Clutch first, and so on, you know the drill. After kangaroo-jumping us on every gear change for about 3 km (2 mi), he started to get the hang of it and we were away. Every time we'd come to a gate or a river crossing, or any time we had to stop, Carmo'd slow down, slower and slower, then I'd scream, 'Clutch!' just as we'd stall.

He did that for the whole nine days. He just kept forgetting. I wish I hadn't argued with my mechanic. (Only joking, Carmo.)

What hat to wear

Wearing a baseball cap in the Aussie sun is about as useful as one walkie-talkie — it just doesn't do the job. There's a good reason why a bloke in the bush wears a hat with a brim — it keeps the sun off his ears and neck, and covers up his melon.

You need an air gap between the top of the hat and the top of your scone. The hotter the place you live or work in, the higher the hat. That's why in the Northern Territory they've got massive big hats, the old 10-gallon jobs. The bigger gap between the hat and the head keeps the brain cooler. The further south you go, the smaller the hat, and the smaller the gap.

I guarantee if you wear an Akubra-style hat out on a stinking hot day, you'll feel 50 per cent better than if you wear a baseball cap. Try it out.

I've had lots of young blokes work for me. The first day one of them wears a baseball cap, he feels crook — bit of sunstroke, the whole shebang. Second day, I bring in a spare hat (I've got about 10 good hats).

Baseball caps are for baseball, but if you're gonna wear one, don't wear it backwards — that's for the Yanks. Let's stick to Aussie gear.

Another tip — it seems pretty obvious, but keep the fluids up. Drink heaps of water. The way to do it is to cart around a big foam drink container. That way you'll remember to have a regular drink. If you're drinking when you're really thirsty and hot it's too late, you're already dehydrated.

It's happened to me. I was a young bloke working in South Australia — not enough fluids throughout the day and all of a sudden dehydrated, bit of sunstroke. She was a beautiful 43°C (110°F) Adelaide day, so I started pumping water in to right myself. But it was too late, the damage was done. I kept gulping water but I still wasn't feeling any better. Then I started spewing, I was crook. The boss had to let me go home. He wasn't very happy.

Remember: a good hat, plenty of water, and at 5 pm, a cold beer.

THE CAR

These days working on the car is almost a no-no. With the computers and technical stuff under the bonnet, you almost need a four-year tech course under your belt before you can even put fuel in it.

Don't let them beat you. You can still do an oil change, replace the oil filter and do a grease change. I also like to have all the mechanical tools in the shed. I've got an old Falcon I do a bit to, and I love the mechanic's tools.

Changing the oil

Underneath your car is the sump, which holds all the oil. You need to drain all the old oil from the sump, then replace it with fresh oil.

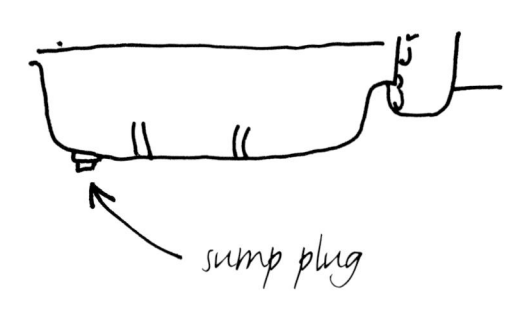

sump plug

1 In the centre of the sump is a sump plug. Of all the bits in your motor, it should be the closest thing to the ground. Get yourself a bucket so that you can drain the oil into it. Under no circumstances should you let the oil go down a drain or on the path or driveway.

2 Undo your sump plug. Let the oil drain, then drop in about half a litre (17 fl oz) of new oil that'll flush out the rest of the black oil.

3 Replace the sump plug and fill up with oil.

Changing the Oil Filter

Now replace the oil filter. You may need a special tool to undo the filter. You can pick these up at an auto spare parts shop. The oil filter is a cylinder with writing all over it. And it will be labelled 'oil filter'.

1 Undo the filter. It just screws out. Take this down to the auto spares shop and get a replacement.

2 Replace it, then screw it back in.

An old mechanic once said to me, 'If you do an oil change on your car every three months, you'll get a heap more life out of your motor.'

Changing the Grease

Another little piece of maintenance that can be done on the car is the old grease change. You'll need to buy a grease gun, and of course some grease.

1 The front end of the car has the most grease nipples. Yes, nipples. That's what you're looking for. If you're not sure what's what, next time you get fuel, ask the mechanic what to look for.

2 Once you've established where they are, just get underneath. Clip the end of the grease gun onto each nipple. Give the gun a couple of pumps, until a little excess grease pops out. The job's done.

There's nothing finer than doing a little bit of work on the car. If you want to go a bit further, just about every car has a DIY-style maintenance book that suits

THAT CAR ONLY. I'VE BOUGHT THEM OVER THE YEARS FOR ALL MY OLD CARS, WHICH IS A STORY IN ITSELF.

TIP OF THE DAY: COLLECT BRICKS

I USED TO WORK FOR AN OLD BLOKE WHO TOLD ME, 'NEVER THROW AWAY A BRICK.' HE USED TO PICK UP SINGLE BRICKS, WHACK 'EM IN THE UTE AND ADD TO THE STACK AT HOME. IT'S A GOOD POINT: ONE BRICK IS JUST A BRICK, BUT MANY BRICKS MAKE A WALL, AND MANY WALLS MAKE A HOUSE.

SO THESE DAYS I NEVER THROW AWAY SINGLE BRICKS. IF I SEE ONE, I PICK IT UP AND SAY TO MYSELF, 'YOU CAN BUILD HOUSES OUT OF THESE BABIES.'

MULCHING

Forget weeding — that's for knuckleheads. Mulching is the way to go. All your pots and garden beds should always be mulched. There are heaps of different varieties and different costs — that's up to you.

Why mulch?

1 It looks good, and finishes off your garden beds and pots.

2 It retains the moisture in the soil. Less water is lost through evaporation, and more water is absorbed by the soil.

3 You use less water. Good for the environment and good for your wallet.

4 It suppresses weeds, so there is less work for you to do.

Stop arguing with me and mulch!

THE EVOLUTION OF THE FLANNO SHIRT

New flanno, straight out of the packet. Wear it in winter till the fabric thins out a bit.

Once your flanno gets a little worn, it becomes light enough to wear in summer. Just cut the sleeves off and you're in business.

Use the cut-off sleeves as rags.

And recycle the cuffs by using them as stubby holders.

121

Enjoying

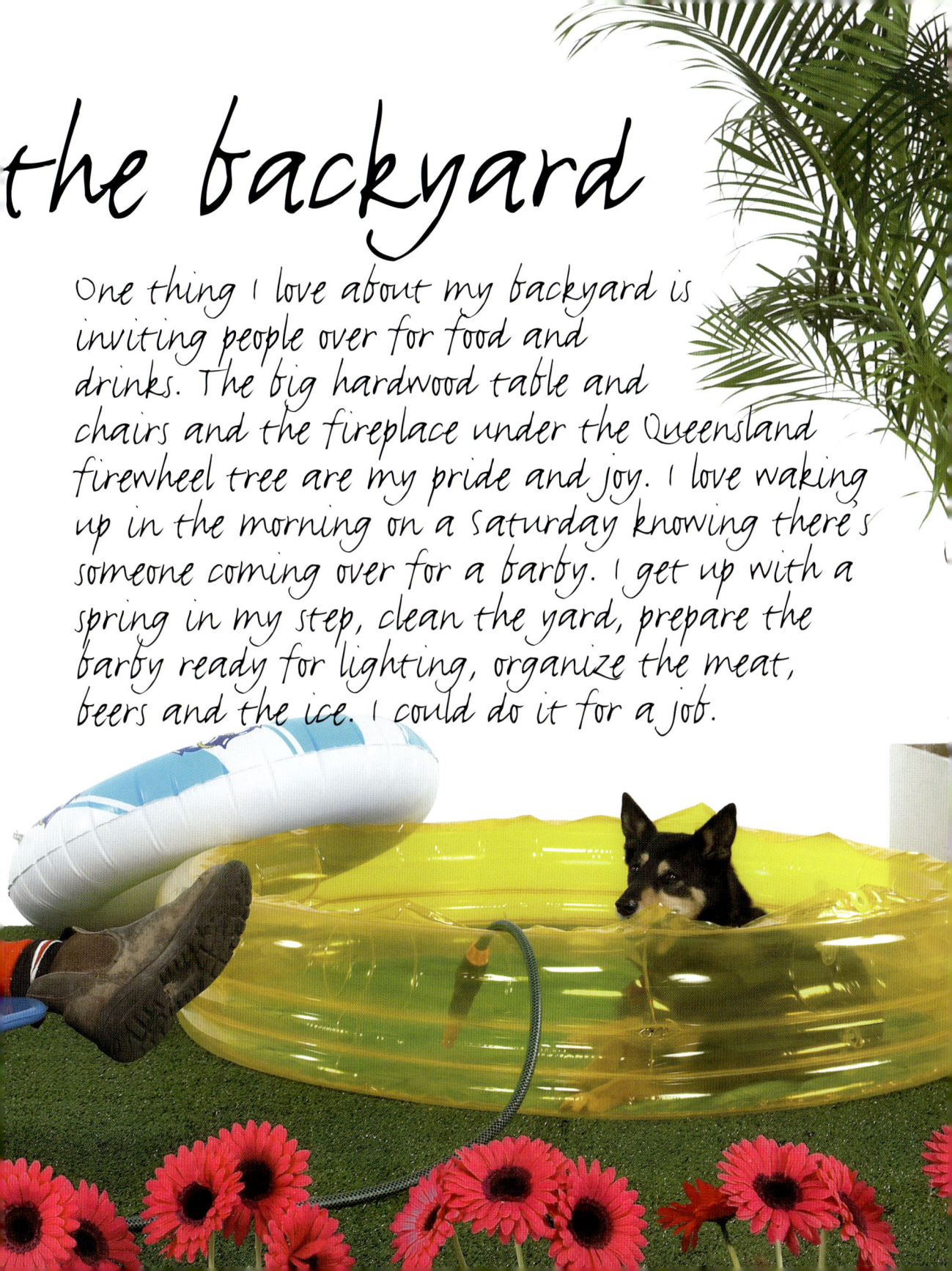

the backyard

One thing I love about my backyard is inviting people over for food and drinks. The big hardwood table and chairs and the fireplace under the Queensland firewheel tree are my pride and joy. I love waking up in the morning on a Saturday knowing there's someone coming over for a barby. I get up with a spring in my step, clean the yard, prepare the barby ready for lighting, organize the meat, beers and the ice. I could do it for a job.

THE BARBY

To me the only way to cook everything is on the barby. I've got the open-fire barby set up beneath a 60-foot tree, you should see it. It's like something straight out of 'The Man from Snowy River'.

If I need a quick cook-up, I've also got the world's biggest gas barby. This thing is so huge you could sleep in it.

The barbies in the backyard are my favourite things. On weekends I cook three meals a day on the barby. It's a great Aussie pastime.

Cooking on the open fire

My grandfather used to cook snags on the open fire, with the flames leaping out of the fire pit, fuelled by the fat squirting out the sides of the snags. They'd come off the grill black as, and he reckoned the charcoal was good for your teeth.

It's not a good idea to follow my grandfather's lead and burn all your food, even though charcoal is good for your teeth.

The key to not burning food is not to have a flame; your fuel in the fire has to burn out, leaving red-hot coals to cook on. There must be enough hot coals in the fire to last the cooking time, so you really need to get the fire going about one hour before you cook. Build it up as big as your barby can take it without being ridiculous, and leave it until you're left with red-hot coals. Now you're ready to cook.

And have a shovel handy to shift coals around to adjust the heat in various parts of the plate. If you can handle cooking on the open fire for 10 people, consider yourself a champ.

TIP OF THE DAY: FOLDING GRILL
GET YOURSELF A FOLDING GRILL WITH A HANDLE. IT SITS STRAIGHT ON TOP OF THE GRILL PLATE AND YOU CAN TURN EVERYTHING IN ONE HIT.

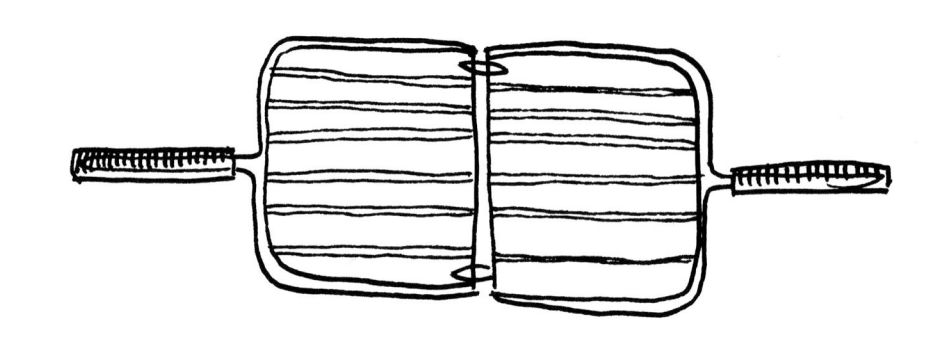

TIP OF THE DAY: WET TIMBER
STARTING A FIRE BUT THERE'S ONLY SOAKING WET TIMBER TO HAND? IT CAN BE DONE. SPLIT THE WET TIMBER WITH AN AXE. THE CENTRE WILL BE DRY. SPLIT THE DRY TIMBER UNTIL YOU HAVE 15 OR SO DRY PIECES. EACH OFFCUT WILL HAVE AT LEAST ONE DRY FACE. BUILD THE FIRE WITH YOUR DRY KINDLING, THEN BUILD THE FIRE UP WITH THE OFFCUTS, DRY SIDE DOWN ON THE FLAME. SPLIT YOUR BIGGER LOGS AND KEEP THEM CLOSE TO THE FIRE, AND WHACK THEM ON AS THEY DRY.

BUILDING THE FIRE

This is the only way to build a fire. My father taught me this when I was about five years old. It's foolproof and you'll get a crackerjack fire going in about two minutes. All the measurements given for the timber are across the face. I've taught my eldest son. He's an expert fire builder.

1 All your timber should be about 300 mm (12 in) long. Start with two 100 mm (4 in) logs, and place them approximately 300 mm (12 in) apart.

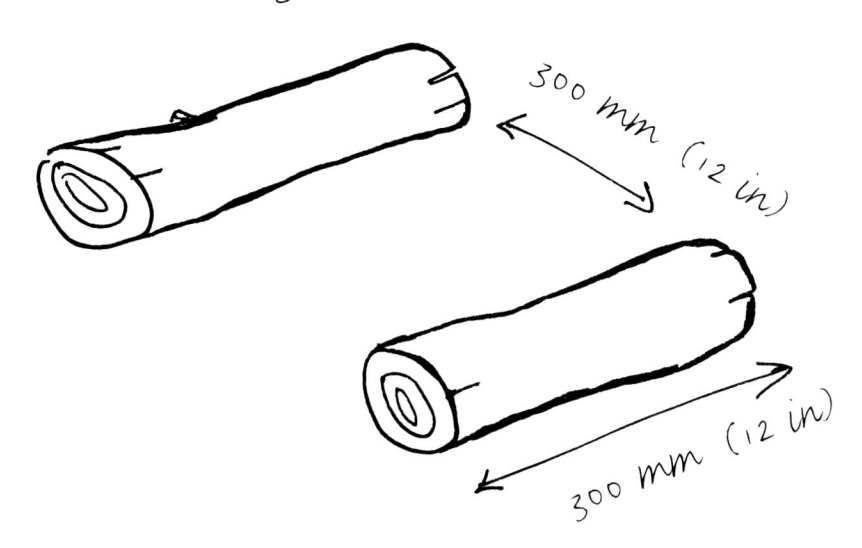

2 Scrunch up 10 sheets of newspaper reasonably tight, and toss them in between the bare logs.

3 Kindling is next. It should be hardwood. The best kindling I know, 'cause it's easy to get and free, is old fence palings. You'll often see them on the side of the road or at building sites. Grab 10 or 15 hardwood palings, cut them up to 300 mm (12 in) lengths, and then split them with a tomahawk into approximately 10–15 mm ($3/8$–$5/8$ in) pieces. The old palings split really easily, and in no time you've got a

WHOLE PILE OF KINDLING TO LAST THE WINTER. LAY SIX OR SEVEN PIECES, BRIDGING THE TWO BARE LOGS, THEN SIX OR SEVEN THE OPPOSITE WAY ON TOP.

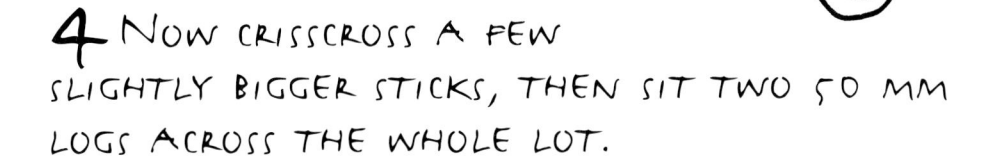

4 NOW CRISSCROSS A FEW SLIGHTLY BIGGER STICKS, THEN SIT TWO 50 MM LOGS ACROSS THE WHOLE LOT.

IT'S IMPORTANT TO LEAVE A GAP BETWEEN EACH PIECE YOU LAY DOWN. THAT WAY THE AIR CAN FLOW BETWEEN ALL THE PIECES, AND THE FLAME CAN LIGHT THE PAPER IN THE FOUR CORNERS. YOU CAN THEN SIT BACK AND WATCH THE FIRE TAKE OFF. ONCE IT'S GOING, YOU HAVE TO KEEP IT GOING, SO STOKE THE FIRE WITH 40-50 MM (1½-2 IN) STICKS, TWO OR THREE AT A TIME. BEFORE ANY MORE GOES ON, MAKE SURE THE PREVIOUS two are well alight and established. Once you've got a good solid base of hot coals and solid flame, a couple of 100 mm (4 in) hardwood logs can go on.

If the big stuff goes on too early, before the fire is established, it'll go out.

TIPS OF THE DAY: OPEN FIRES

1 Never use treated pine to burn in a fire, as it gives off a poisonous gas. You'd have a barby all right, but you'd kill all your mates.

2 Always check the fire ban situation. If you're allowed to light a fire in the bush, make sure the fire is well and truly out before you leave. The best way to do this is to shovel soil or sand on to the whole fire and bury it.

Camping with my little mate

Me and my son Charlie go camping in the bush together. I've got a rough bush block up at Mudgee in country NSW, and we swag out up there on the top of the mountain — no tents out under the stars.

This block of mine is a bush rock farm. Twenty-two bull ants to the acre — you couldn't grow chokoes on this joint. Thick bush, cliffs, snakes, rocks, poor soil, steep hills and bloody cold — paradise. I love it. The bloke I bought it off was trying to sell it for about two years. I think he nearly had a stroke when I said I'd take it. The only thing missing is the power lines.

I know it's a bit of a take from the Aussie movie — but I did have this joint before 'The Castle' came out. Ah, the serenity...

I love getting the kids up there, and walking around, building things, having a few beers. There is no power, no hot water and no cold water. You can imagine what the Minister for Clips Behind the Ear thinks of it — she loves it.

So me and my little mate were about to go on our way one weekend when we got news that Mudgee had a cold snap coming through and it was going to be −10°C (14°F). A bit chilly on the 'Camdarosa' mountain. So I said to the Minister for Stating the Obvious, after she reckoned we'd need extra jumpers: 'Pack us a couple of extra jumpers.'

Charlie Boy and I had a couple of bags that would've been at home on the carousel at the international terminal. We looked like we were going on the three-month Ashes tour.

After a terrific drive up and a great day just doing stuff, we bedded down for the night in our respective swags. Sure enough, it was about −8°C (18°F). Charlie was about six years old at the time and loving it. He had on a beanie, gloves, scarf, socks and boots — normal clothes. If you're in a good sleeping bag and a swag you'd be surprised how warm you are. A nice fire the size of a VW Beetle helps as well.

Charlie and I were up there for three days and two nights. We got a little dirty, and loved every minute of it.

When we arrived home the Minister was horrified by our condition — dirty hands and face (to me there's nothing finer) — but she was even more horrified when she realized we had the same kit on that we'd left in and hadn't so much as unzipped the tour bags.

What a weekend!

Cooking a Lamb Roast in the Camp Oven

Just try this camp oven recipe and you'll be the envy of all your friends. You'll never look back. This will be the best feed you'll ever have, because:

1 You've cooked it Aussie bush-style right there in front of your family or friends' amazed eyes.

2 It's the best way to cook a leg of lamb — it's so tender, it breaks away in flakes and melts in your mouth.

3 Besides lamb, my second favourite food group is spuds — and these spuds are better than ya Mum's. Now I know that's a big call. I can almost hear everyone saying, 'Are you bagging my Mum's spuds? I'll knock you out for that!' Well, I'd never bag ya Mum's anything.

What you need

1 camp oven (disposal store)
1 long-handle shovel (hardware)
or
1 long stick (bush)
1 boneless leg of lamb (butcher)
Spuds and pumpkin (vegie shop)
All-purpose seasoning
Rosemary
Salt and pepper
1 bloody hot fire (anywhere you're allowed to have one)

Preparing the hot coals

This part is tricky, but the most fun. You need to prepare the fire about five hours before you plan to eat. If you haven't got an open-fire barby in the backyard, you'll need a patch of grass about 600 x 600 mm (2 x 2 ft) out in the middle of the lawn.

1 Just lift the turf up with a spade, put it aside, water and cover it. The turf will look nearly the same when you replace it, and in a couple of weeks it'll be back to normal.

2 Dig out the soil to about 150 mm (6 in) deep, keeping the soil in buckets or a wheelbarrow to make replacing it easy.

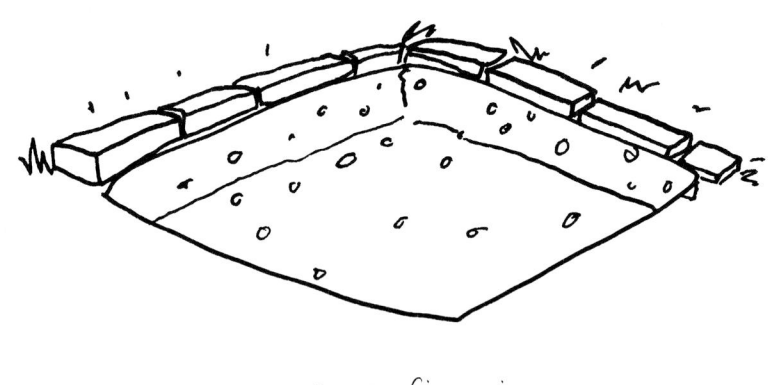

Open-fire pit

Cross-section of open-fire pit

3 If you have a few bricks handy, line the edges to stop more lawn burning than necessary. A good hardwood fire is the go (see the instructions for 'Building the fire' on page 127).

4 Keep stoking it for about two hours, then let it burn for about 45 minutes, basically until you've got bloody red-hot coals about 200 mm (8 in) deep (these measurements are all rough). You need a really screaming-hot pile of glowing coals.

5 OK, so two hours before you want to eat, prepare the lamb and vegies.

Preparing the roast

1 Get your large camp oven and throw a dash of oil in, just enough to cover the bottom. I sometimes throw about 5 mm (¼ in) of red wine in the bottom of the pot for extra flavour.

2 Throw in the lamb, keeping it in the middle, then throw in the spuds and pumpkin (peeled if you prefer), a bit of salt and pepper, all-purpose seasoning, rosemary...then whack the lid on. (Did I mention this is perfect for a bloke, because prep time is about five minutes?) OK, now you're back out to the fire.

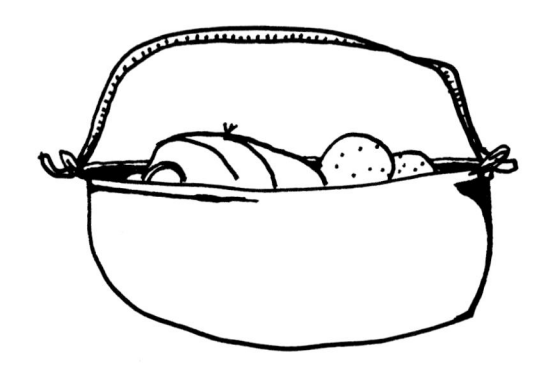

Camp oven

3 Use a long-handle shovel to make a depression in the coals roughly the size of your camp oven.

4 With a long stick or something similar, hook into the handle of the camp oven and place it in the depression you've made in the coals. Your friends and family will now be looking at you in amazement — at this point you are an official legend in their eyes, and there's more to come. Using the long-handle shovel, move some of the coals from the edge of the fire and cover the lid of the camp oven.

Cross-section of camp oven in coals

5 Right, now the hard bit. Stand back with a couple of cold cans for 1½ hours and watch the fire.

6 Remember to slowly keep the wood up to the fire, but keep it out towards the perimeter of the pit. Depending on the

HEAT OF THE FIRE AND THE SIZE OF THE LAMB, IT SHOULD TAKE ABOUT 1½ HOURS TO COOK.

7 AFTER ABOUT ONE AND A BIT HOURS, SCRAPE THE COALS OFF THE LID WITH A BIT OF WIRE OR HOOK (JUST MAKE SOMETHING UP — A LONG STICK WITH A POINTY NOSE WILL DO) AND HOOK ONTO THE HANDLE ON THE LID ONLY. DON'T PULL THE OVEN OUT, JUST TAKE THE LID OFF, AND CHECK THE LAMB. (DO THIS THROUGHOUT THE COOKING PROCESS.)

8 IF IT'S NOT READY, WHACK THE LID BACK ON AND COVER THE CAMP OVEN WITH SOME FRESH COALS. THEN LEAVE IT UNTIL IT IS COOKED.

THROUGHOUT THE COOKING PROCESS, YOU MUST KEEP SHOVELLING IN FRESH COALS AROUND THE LID OF THE OVEN. THESE COALS COOL OFF QUITE QUICKLY — JUST

Pinch hot ones from around the edge every now and then.

The whole concept of this cook-up is improvization. Use whatever you can find, and when someone says you can't do that in the backyard, you say, 'Why not?'

Always have the hose handy, and check for fire bans and the rules at the time for fires.

Replacing the turf

Next day, after everyone stops patting you on the back, take three-quarters of the ash out of the hole. Mix the soil in with the remaining quantity of ash to the right level, then pack it down. Replace the turf, give that area a good water every day for a week, and no one will ever know you had a fire there. What a feed!

Open-fire barbies

There are a few ways to attack the open-fire barby. At one time or another, I've had all of them.

Two-stone barby

The easiest, and also my favourite, is just two sandstone blocks with an old barby plate sitting over the top — too easy. Dig the two rocks into the ground a bit, just to level them up.

Two-stone barby

I'm a bit of a scrounger! I drive around the streets with one eye on the road and the other on the nature strip, looking for stuff — anything that looks useful. Then I usually hang on to it for a couple of years before I toss it.

One of my obsessions is barby plates. I have about 50 of them. I usually get them from old barbies that have been tossed. If I'm driving along and see a barby on the side of the road, I'm straight onto it. A lot of the time the plates are gone, someone has beaten me to it, which is good news for me! It means there's some other bloke out there with 50 plates. There's at least two of us.

These plates, or grills, can be used on the two-stone barby. If you're doing brekky, whack on the hot plate for bacon, eggs and baked beans. When you're doing lunch with the T-bones or chops, on goes the grill.

The grill plates can stay on the stones, and another set can be wrapped in an old towel and kept in the boot of your car, so if you're ever in the bush or on a picnic, just find two stones or bush rocks, and whack the plates on top for an instant barby.

Or, you could go extremely flash as I did, and have a three-stone open fire. That way you can have a grill on one side and a plate on the other.

Three-stone open fire

My barby

My open-fire barby is one step up from the two-stone fire, although it started out as a two-stone before it was elevated into a magnificent piece of equipment that I love like a fourth child.

My three-stone open fire

This type of barby is easy to build, and I reckon it looks great. I just collected a pile of sandstone out of skip bins from various jobs. Every time I saw an excavation site, I'd stop and see what I could find. When I had a big enough pile, I started.

Basically then, it's just a bit of a jigsaw puzzle. It took me about three goes, figuring out which stone to start with, and I changed the design a few times along the way. No two fireplaces look the same, as there is no plan. The design is in your head and comes out as you build it. And the bottom line is, if it cooks a snag, then it works.

Brick Barby

The same principle can be applied to bricks. Lay them dry, straight on the ground, although it's important to start off level and every so often just straighten up the bricks. Nothing worse than the grill rolling in the fire with all the snags on board.

Damper on a stick

This recipe is a cracker for the kids, or even for your barby guests.

1 cup self-raising flour

Pinch salt

Pinch sugar

½ cup beer

1 Make up a damper mix with the flour, salt and sugar, using beer instead of water to mix it up into a nice dough.

2 Grab a handful of dough about the size of a fat sausage, and mould it on to the end of a long stick or a thick bit of wire, then hold it over some bright red coals — not flame.

3 Cook the damper for about five minutes, but keep checking it in the meantime. When it's finished, you've got yourself a fresh damper roll. Whack in a bit of butter — you beauty!

Barby etiquette

There are a lot of myths about the barby. My old mate Geoff Janz reckons you should only turn a steak once. Well, that might be right when you're a chef but, fair dinkum, Janzy.

The whole thing about a barby is turning the snags and T-bones, tongs in one hand, beer in the other, flames leaping out of the grill and singeing the arm hairs. Turn the steaks 50, maybe 60 times, roll the snags 100 times. If you've got a pair of tongs, get 'em in there. Idle tongs are for chefs. Janzy's gotta blow the dust off his tongs.

And another thing. I always have my barby on low. If it's up too high, it's over all too quickly. Keep it on low and make the experience last longer. I'm always a bit upset as the last snag comes off the grill — I'm looking at an empty barby plate, tongs in hand, a tear in my eye.

That's why, if you've got a fancy barby, it's a good option to put the rotisserie on with the lump of beef or the boneless leg of lamb with the hood down. They take two hours and in front of your mates you can check it maybe 250 times.

As you open the lid, everyone goes 'Whoaaaaa... beautiful!' — and you say in reply, 'Just a little bit longer.'

I pride myself on being good on the barby. Once I cooked seafood for 40 people on the open-fire barby, turning the fish and prawns, and stoking the fire at the same time. People starting calling me Peter Doyle after that one...

Although my organization was impeccable on that particular occasion, I think I might have sent the Missus into the kitchen for a tray or plate maybe 20 times. And I always forget a sharp knife...

TIP OF THE DAY: PERFECT SNAGS
Cooking sausages to perfection is pretty simple — put them on a low heat, and cook them slow and long. If you do it that way, the outside won't burn and the inside will cook through.

If you're on the grill and the flames are coming up, move the snags around until the oil stops burning.

If you're on the hot plate, roll the snags in flour. It gives them a bit of a crispy coating. They're sensational.

And you'd have to be born on Mars not to know to prick your snags with a fork as they're cooking. If you don't, they'll burst. Bog in!

THE BAR

The home bar — every bloke's dream. Draught beer at home, your own on tap. Obviously, it's out of reach for most of us, and even if you can afford it, the likelihood of the Minister allowing it to slip through the net would be about 1000 to 1. The home bar can take any form really, from the replica of your local to two Eskies full of ice sitting under a trestle table.

At the end of the day, the bar needs to accomplish two things:

1 plenty of beer, and

2 keeping it cold.

Pretty simple, really.

The Keg at Home

How many times have you been to someone's house and they've whacked on a keg, but then the thing starts playing up? It's all heady, so you start pouring jugs off, but there's wasted beer all over the place. The beer goes warm while it's sitting in the jug and, of course, everyone is an expert. 'Turn the gas down! Turn it up!' It's a bloody nightmare.

The word is, your mate's having a party and he's got a keg on, you just know something is gonna go wrong.

I own a magic box, a beer-chilling device that you can move around. It uses bagged ice to cool the beer so you can have keg beer anywhere — camping, the backyard. It's just fantastic.

The magic box is normally what's used in the backyard keg set-up. If you know the principle behind it all, you've got half a chance of fixing it and being a hero.

magic box

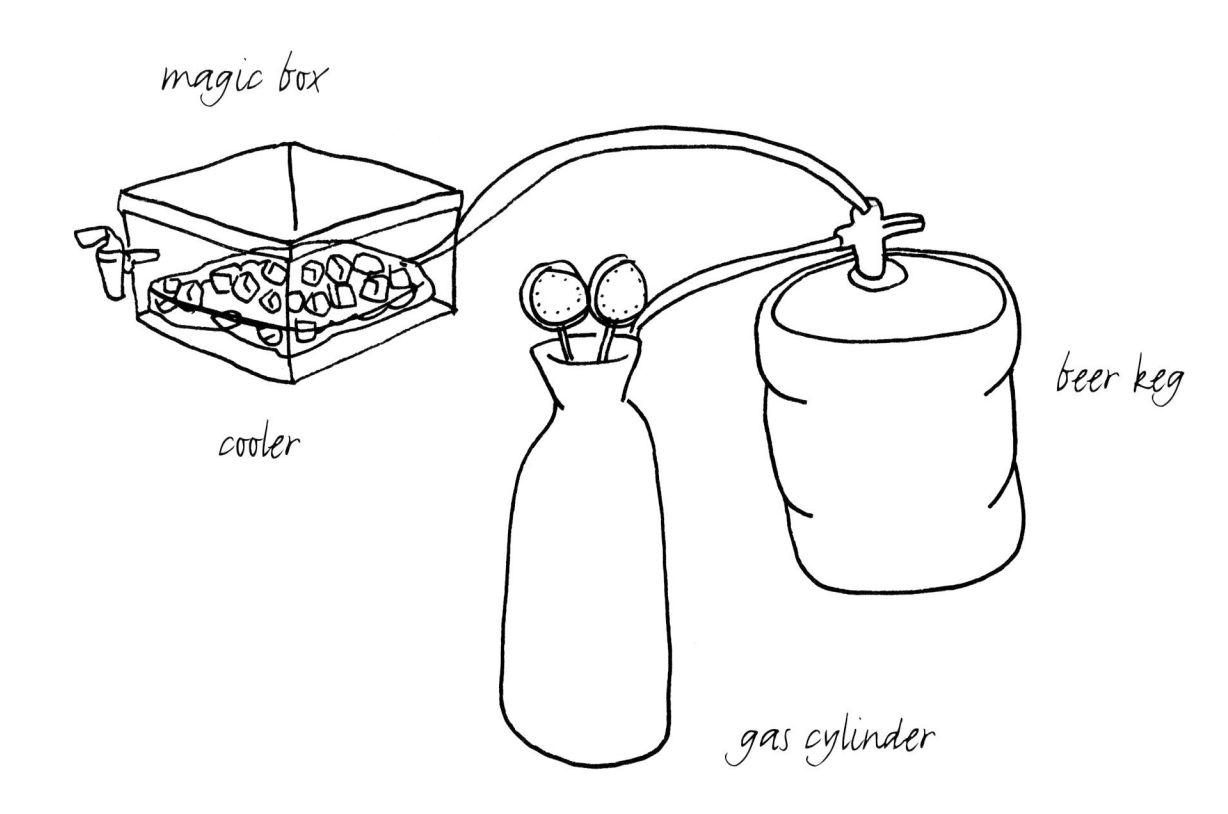

cooler

beer keg

gas cylinder

to the gas

to the tap

THE BASIC IDEA IS THAT THE KEG NEEDS TO BE UNDER PRESSURE, OTHERWISE IT BUBBLES UP, CAUSING TOO MUCH HEAD, SO TURNING THE GAS DOWN WILL ONLY MAKE IT WORSE.

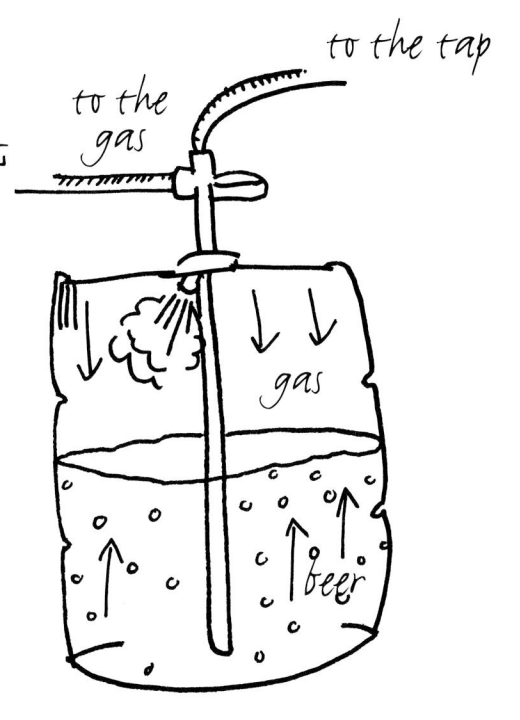

gas

beer

So the beer is always under pressure in the keg. When the tap at the magic box is opened, the pressure from the gas, which is constantly pushing down, forces beer from the bottom of the keg, up through the tube inside the keg, and out down the lines to your glass.

As soon as the beer hits the outside world, the bubbles start to rise. If there's not enough gas in the keg — which means not enough pressure — the beer will head up in the keg.

If the keg is pouring badly, there are a few possible solutions.

1 Turn the gas up by turning the regulator clockwise. The maximum you want to go to is about 220 lb.

2 Keep the keg cold, as a warm keg will cause the beer to pour badly. Put the keg in the shade, then wrap it in wet towels and lay ice over the top. Keep the towels wet with the hose.

3 Make sure the lines are clean. If it's two o'clock in the arvo and the beer's pouring badly because the lines are dirty, it's something to remember for next time. At the end of the day, don't be frightened to whack the keg on for a party. There's nothing finer than draught beer at home.

TIP OF THE DAY: BOTTLE OPENER

Always open a non-twist-top bottle of beer with anything but a bottle opener. It's very impressive. Any of these items can be used in the same way to open your cold one. Basically, all you need is a solid object, with an edge:

* Disposable lighter
* Butter knife
* Chisel
* File

Grip the bottle neck tightly and keep your fingers up close to the cap. Using your index finger as the leverage point, slide the top edge of whatever you're using as an opener, and pop the lid off. The key to the whole operation is to act casual.

Priming the Pewter

I always love drinking cold beer out of a pewter mug. My Mum gave me two of my late father's favourite pewters. The handle on each of them is a golf bag with clubs (he won them at golf). They keep the best head all the way down, leaving beautiful sip rings down the inside of the pewter.

My father taught me how to keep a beer pewter in perfect condition so it'll always give a good beer head.

First, the ongoing rules of pewter.

1 Wash your pewter in hot water only. Do not use detergent.

2 Milk and fruit juices ruin the inside of a pewter. Only drink beer out of it.

3 Lipstick contaminates the beer and sends it flat quicker.

4 A dirty mug will also contaminate the beer and make it go flat faster.

The way to keep your pewters improving all the time — and, more importantly, to bring back some favourite ones from the dead — is to prime them.

1 Get yourself a couple of cans of beer, line your pewters up on a shelf or somewhere they can't be disturbed, and pour the beer in. Try for no head. Let that go flat a bit, and keep topping up until your pewter is absolutely full of beer, with no head. It's important to get a convex effect on the beer out the top of the mug by just putting in a couple of drops to finish it off.

2 Leave the pewter like this for at least 24 hours (no more than 48 hours), then rinse it out with hot tap water and leave it to drip dry. Repeat this a few times if you're trying to bring one back from the dead. Once it's drinking nicely, prime your pewters every three months. Cheers!

Bar etiquette

There aren't many rules behind the home bar. You've got to make them up as you go along.

At the end of the day, you're the licensee at the home bar, and you make the rules. It goes without saying that you must be fully stocked, and the stock must be cold as buggery. There is absolutely no excuse for either of these being below par.

Don't be fooled into thinking the ice will last on a stinking hot day. No matter how much ice you buy, it won't be enough. The best you can do is keep the eskies in the shade and maybe have three or four spare eskies, just with extra ice as backup. Nothing keeps a beer or wine cold like ice in an esky.

And finally, always buy an extra couple of slabs of beer and extra bottles of wine, so you end up with some grog left over. (As if it'll ever go to waste!)

Nicknames

Aussies are famous for giving out nicknames. I love a good nickname, especially when it evolves into something so obscure you forget where it started.

I know a landscaper we call 'Billions', but not because he's loaded. Years ago, a few of us blokes were working for this woman. Every arvo she said, 'Thank you, backyard people' (with a bit of an accent). Lovely lady, I might add. So we picked up on this a bit, and when someone brought something to you or helped you, you'd reply by saying, 'Thank you, Scott people or Phil people or Mathew people.' For some reason the 'people' bit stuck to Mathew, so he was 'people' for two years. Then one day on holidays, we found out a guy called People was a red-hot ping

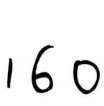

160

pong player — they play ping pong in China and there's billions of people there. Fairly obvious, really.

I also know a bloke they call Esky. When I first met him I thought, 'He's a bit stocky, maybe drinks too much.' So I asked the boys, 'What's the go with Esky?'

'When he gets full of grog, we've gotta carry him.'

One bloke we used to call Fair Dinkum Baz, because he was like a mate of mine, Fair Dinkum Kev. When Kev told a story all you said was, 'Fair dinkum, Kev.' He was in the habit of telling a few pork pies.

Baz started to think 'Fair Dinkum' was a term of endearment, so his name got changed to 'Show Bags', because he was full of shit.

I think the great thing about Australia is how we put a 'y' or 'o' on the end of everything. I don't know anybody who I haven't given a 'y' or 'o' to — Ricko, Carmo, Mungo, Heffo, Poly…and Smoko.

Even my kids started early with 'Mummy'.

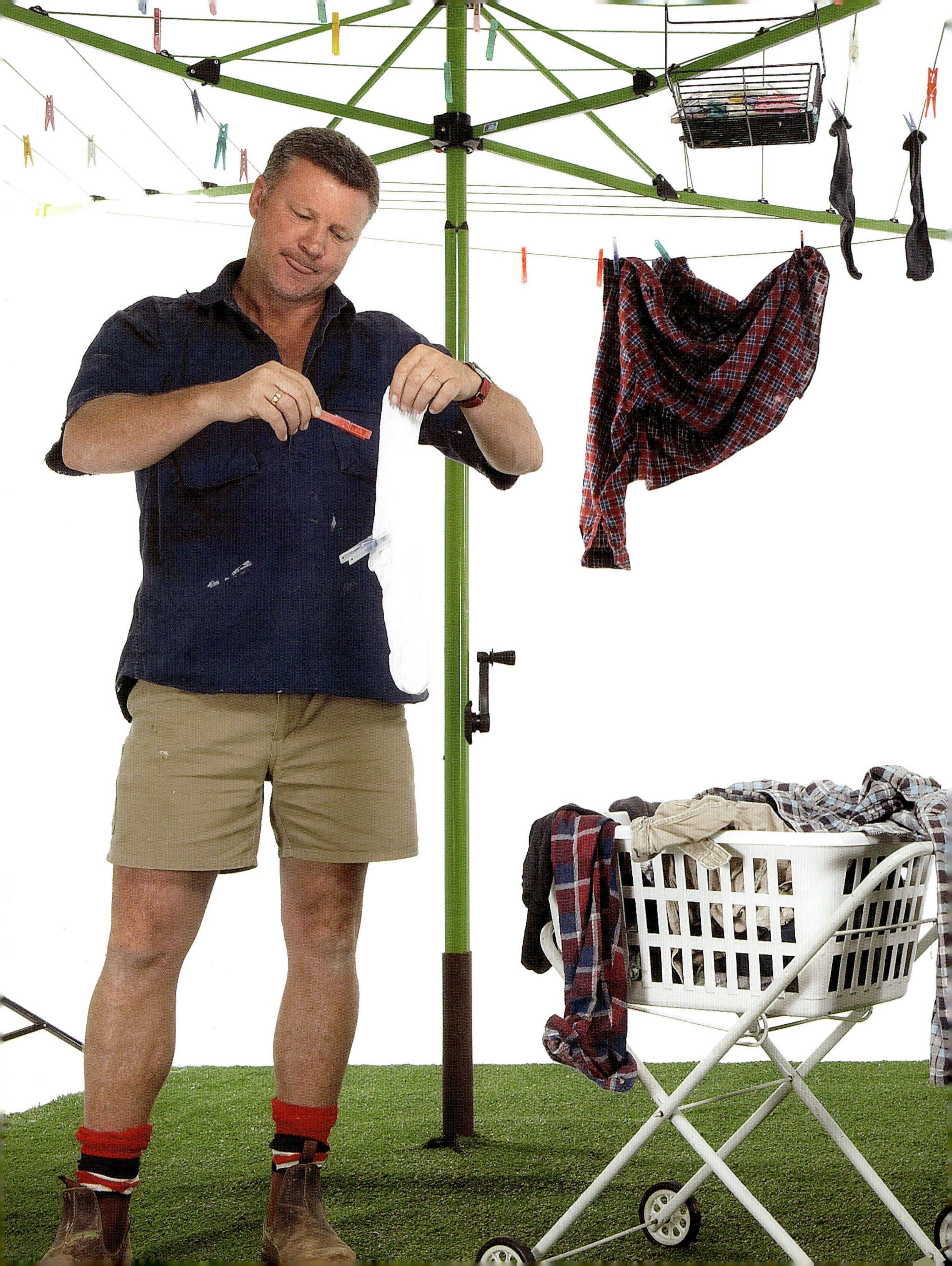

Domestic duties

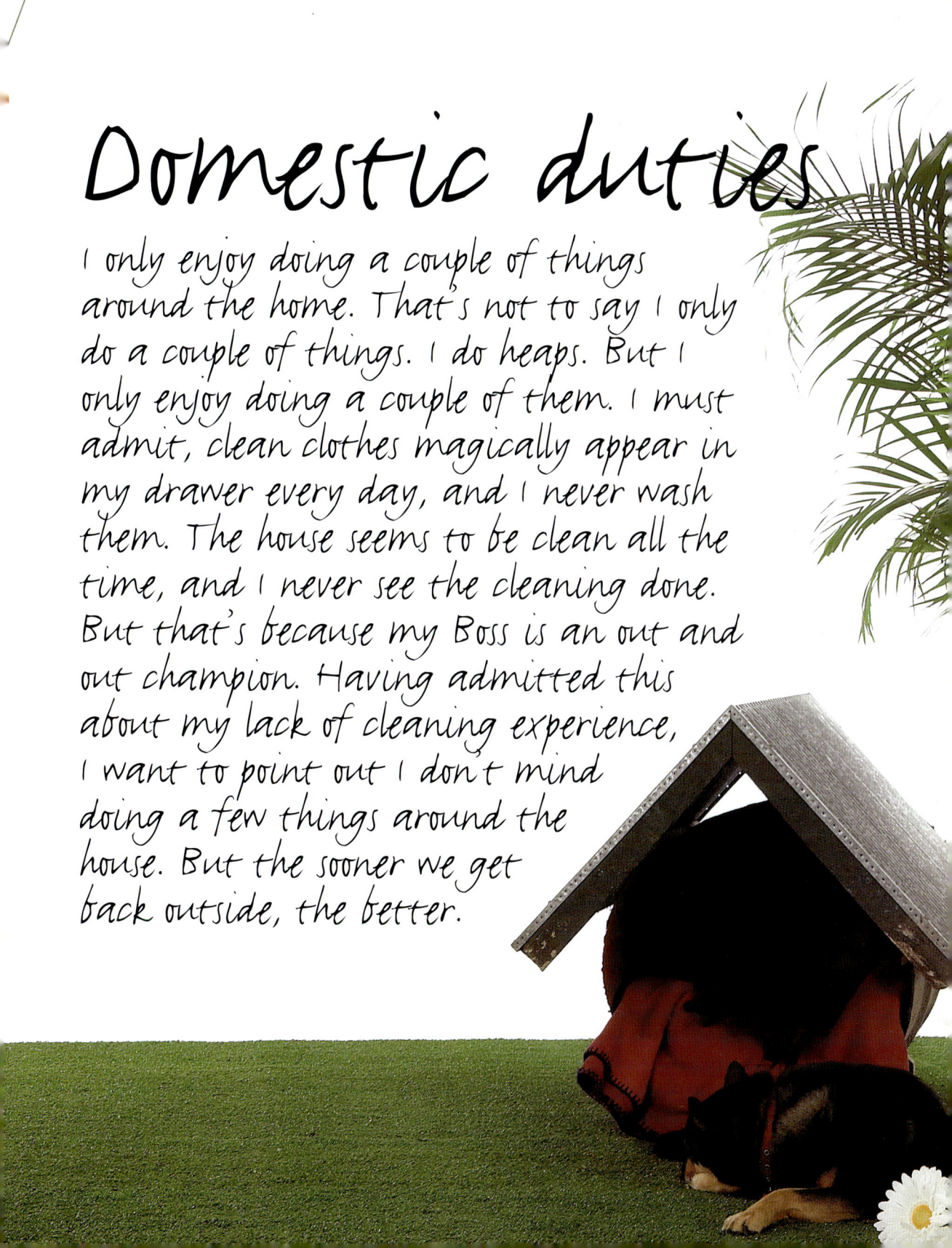

I only enjoy doing a couple of things around the home. That's not to say I only do a couple of things. I do heaps. But I only enjoy doing a couple of them. I must admit, clean clothes magically appear in my drawer every day, and I never wash them. The house seems to be clean all the time, and I never see the cleaning done. But that's because my Boss is an out and out champion. Having admitted this about my lack of cleaning experience, I want to point out I don't mind doing a few things around the house. But the sooner we get back outside, the better.

Dusting

One day, the Minister for Cleanliness asked me and my eldest boy Charlie to dust the whole house — all the shelves, the hall stand...the lot — while she was out shopping.

After 10 minutes of pleading ignorance ('I don't know how to dust', 'I don't want to break anything', 'I have work outside to do') I got the job.

We started the job off fairly slowly. A distinct lack of enthusiasm was the order of the day, until I had a brain snap, I mean, brain wave.

One of the attachments I have for the compressor is a blower. The compressor is a machine that produces high-pressure air, which drives my nail guns. My compressor runs at about 120 psi (that's 120 pounds per square inch). The blower attachment is basically like a small hand-held gun that blows high-pressure air out of a nozzle about the size of a pen.

So Charlie and I got the compressor out of the shed and plugged her into the laundry. We got out the 20 m (65 ft) air hose, which reaches through the whole house, connected the blower up and away we went.

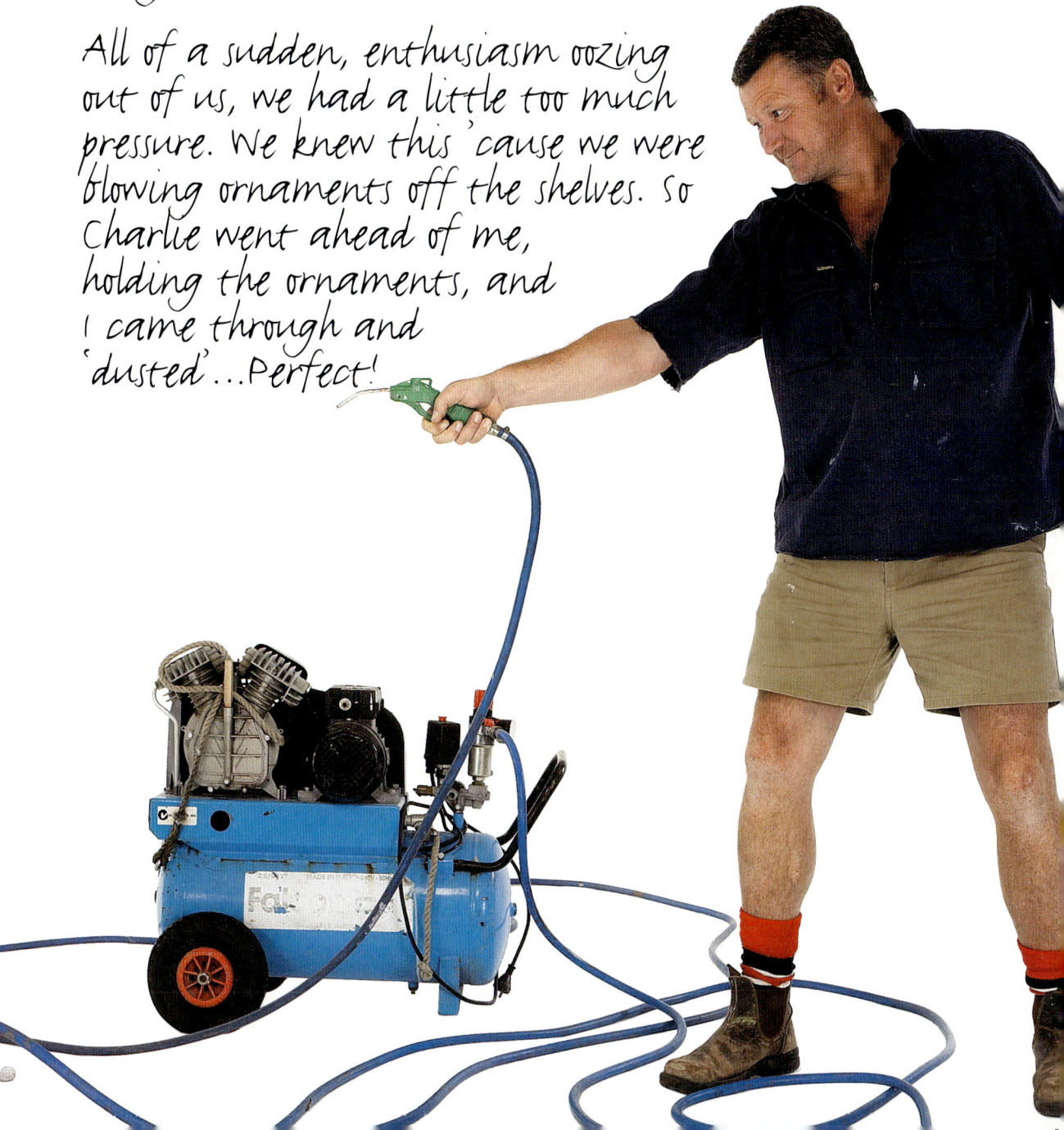

All of a sudden, enthusiasm oozing out of us, we had a little too much pressure. We knew this 'cause we were blowing ornaments off the shelves. So Charlie went ahead of me, holding the ornaments, and I came through and 'dusted'...Perfect!

THE CHOOFER

What's a choofer? It's a hot water system that you can make yourself. Some people call it a donkey. I absolutely love the choofer. It's ingenious yet simple — a 44-gallon drum full of water, heated over an open fire.

Choofers have been used for years in the bush. When I worked in the bush, a lot of stations had donkeys for the boys. I once saw a choofer that handled 20 shearers at a time. But getting the fire going and waiting for the donkey to heat up and it's about $-2°C$ ($28°F$) outside... that's when you're thinking about a proprietary hot water service. But now I'm showering in one of them and I'd love a tub out of the donkey.

If you're planning some renovations to your bathroom and you live in an area where open fires are allowed, rig up one of these. Or take one camping. When we go camping I use a choofer to heat water for the kids and myself.

Materials

44-gallon drum
Metal funnel
3 × 1½ in screwed galvanized elbows
75, 100 and 400 mm (3, 4 and 16 in) bits
  of 1½ in pipe, threaded both ends

Method

Unless you're a welder, don't do this all yourself. You'll need to take the drum to an engineer's shop for the welding.

1 Lay the drum on its side, then use a hole saw to drill a 38 mm (1½ in) hole in the side of the drum, towards the bottom end. Make sure your hole saw is tough enough to go through steel.

2 Now take the drum to an engineer's shop. Ask the welder to weld the metal funnel into the hole you drilled.

3 On the top of the drum are two caps — a big one and a small one. Get the welder to weld a piece of 1½ in pipe,

ABOUT 75 MM (3 IN) LONG WITH THREAD ON THE END, INTO THE LARGER HOLE.

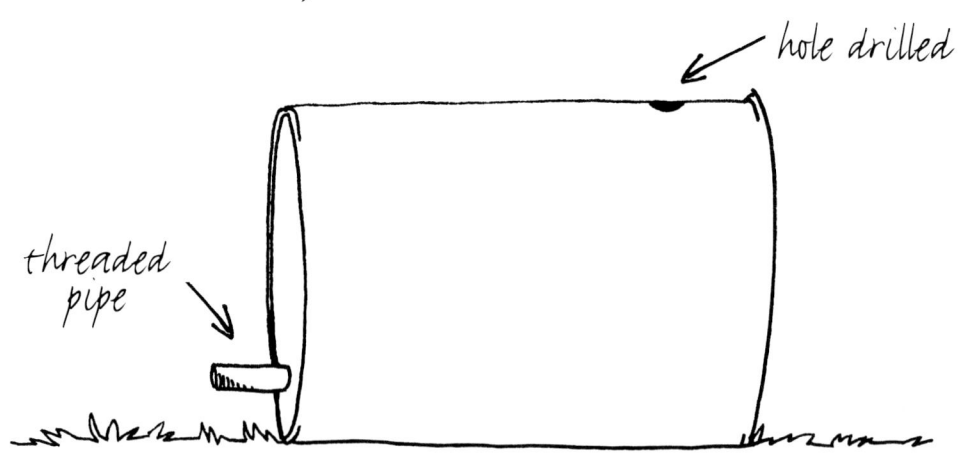

hole drilled

threaded pipe

4 NOW YOU'RE IN BUSINESS. TAKE THE DRUM HOME AGAIN, AND SCREW ON AN ELBOW TO THE THREADED PIPE SO THAT IT FACES UP.

5 SCREW IN A LENGTH OF PIPE ABOUT 400 MM (16 IN) LONG INTO THE PIECE OF ELBOW.

6 NEXT, SCREW IN AN ELBOW FACING AWAY FROM THE DRUM.

7 SCREW IN A 100 MM (4 IN) BIT OF PIPE.

8 Screw in an elbow so it faces down.

9 Get some large rocks or bricks and build a stand for the drum, so that when it's lying down, it's off the ground with room for a fire underneath it.

10 Fill the drum with water and get a nice fire going underneath. After a while, you'll see steam coming from the funnel. You'll need a good fire to heat up the water. Leave it for a while.

11 Now you're ready. Hang a bucket over the threaded pipe, ready for the hot water. For every bucket of cold water poured down the funnel, one bucket of hot water will automatically come out of the threaded pipe. And it'll go all night if you keep the fire going and the water up to it.

The reluctant masseur

About four years ago, my wife gave birth to beautiful twins — a boy and a girl. They are my two angels. Bill is half an hour older than Sarah, and their older brother is Charlie. The kids are the only reason I go to work — otherwise I'd head up the coast in an old bus, fish, have a few beers, do a bit of handyman work a couple of days a week for food, pub and fuel money. (Did I say that out loud?)

Anyway, back to the twins. After Charlie was born, I was a one-time veteran at witnessing birth, so I thought, 'Second time round will be a walk in the park.' But I forgot about the twin factor.

Once Bill was pushed out and I had been attacked verbally by the Boss, it was all relief. Then the doctor opened his trap and said, 'Push again.' I think for a split second we'd all forgotten about Sarah. My Boss is a strong woman (a few people are frightened of her), so when this bloke told her to push again, I thought, 'This could be the last thing he does.' But instead of tearing the doc's crackers out, Ann pushed on to pop out Number 2. Bloody good effort.

By the time we were home with the twins, Ann looked like she'd picked up another couple of mates while she was away — two footies hanging off her shoulders, full of milk and absolutely enormous. She made Anna Nicole Smith look like an amateur.

Ann was having a good go at breastfeeding the two of them at once. It was fairly tough going, and she came down with a really crook neck. So every night it was my job to rub pain-relieving cream into her sore neck and sympathize with her about how bad her neck was and how hard her day had been. I'm not doubting how hard she works, and would never put down the job of raising children. (If she ever found out I said that, I'd be a dead man.)

So one arvo I came home buggered. I'd put in a big day on the tools in the sun and was looking forward to sitting down with half a dozen coldies. I was on the couch halfway through the first one when the Boss sat down beside me, wanting a neck massage.

I'd been doing this for about a month, so I wasn't being heartless. I just wasn't in the mood. But the good Aussie bloke I am battled on — she lifted her hair up and I squirted a bit of cream over her back and neck, and started massaging it in.

But I'd forgotten that my hands were fairly dirty with a little grease and oil — standard trade hands, really. (I like my hands being dirty, 'cause you feel as though you're doing something.) My hands are always dirty, until I have a tub. On this particular day I just sat down for that beer, and all of a sudden I was into the cream.

A little known fact, which I reckon the mob from the cream company don't know: it's the best hand cleaner I've ever seen, like the stuff mechanics use.

I was halfway through the job when I realized I'd been rubbing jet-black cream into the Boss's back. It looked like I'd been rubbing in a handful of shoe polish. Two options: tell her, but then I'd have to wipe it all off and do it again, or keep rubbing until it's all in, tap her on the back and say, 'All done, darling. And while you're up can you grab us a fresh beer? Thanks.'

Facial hair

In the old days everyone had a beard and a moustache. In the bloke department that is, with the odd unfortunate exception. That was the look. Well, what happened? Now we've all got to shave every day.

I'm going all right. I only shave once a week — some poor blokes like my Italiano mate Vinnie shave twice a day. Why, all of a sudden, does society say that to look respectable and clean-cut, we must be shaved?

I'll tell you why, and I know I'm going to get in strife. I can feel it, and I haven't even written it yet. It's the girls' fault. That's right. Women are to blame for the demise of facial hair. The gay community have adopted the mo as a bit of a symbol — good luck to them — but do you know why they are allowed to embrace facial hair? There's not a lot of women hanging around them telling them to shave.

Slowly, over the years, women have undermined facial hair. Complaining how it scratched them, how it doesn't look good — 'Have a shave!' I will admit that I had a mo for quite a few years.

After a constant barrage of nagging for it to come off, I started to wonder, 'What would it be like without the mo?' The Missus could smell the self-doubt and started pounding me with anti-mo talk. She wouldn't let up.

Then all of a sudden I'm there. I'm in the dunny with the foam out, the razor going, then silence — as I stared at myself without a mo.

It didn't stop there. The Boss kept reinforcing how bad and ugly that mo had been, reminding me that if it ever returned, there'd be strife.

It's time to make a stand and bring back the mo. There are a few strong ones out there with the full beard. I look at those blokes with a tear in my eye. They've got what it takes...

Just start with the three- or four-day growth. I've always shaved about every five to seven days. A bit of growth protects my face from the sun and the wind. This is the new mask of the blokosexual...

United we stand, divided we get a clip from the Missus. So stop shaving every day!

Table of Knowledge

Every now and then the Table of Knowledge goes into a bit of a culinary frenzy, and we all think we're Geoff Janz or Jamie Oliver, going over the best feeds — how good we are at making such and such. Someone else is likely to pipe up and say, 'That's crap. You don't put oil in that.' I'm not innocent in this situation, I'm arguing with the best of them. So this section is about giving the knuckleheads a go in the kitchen. Enjoy.

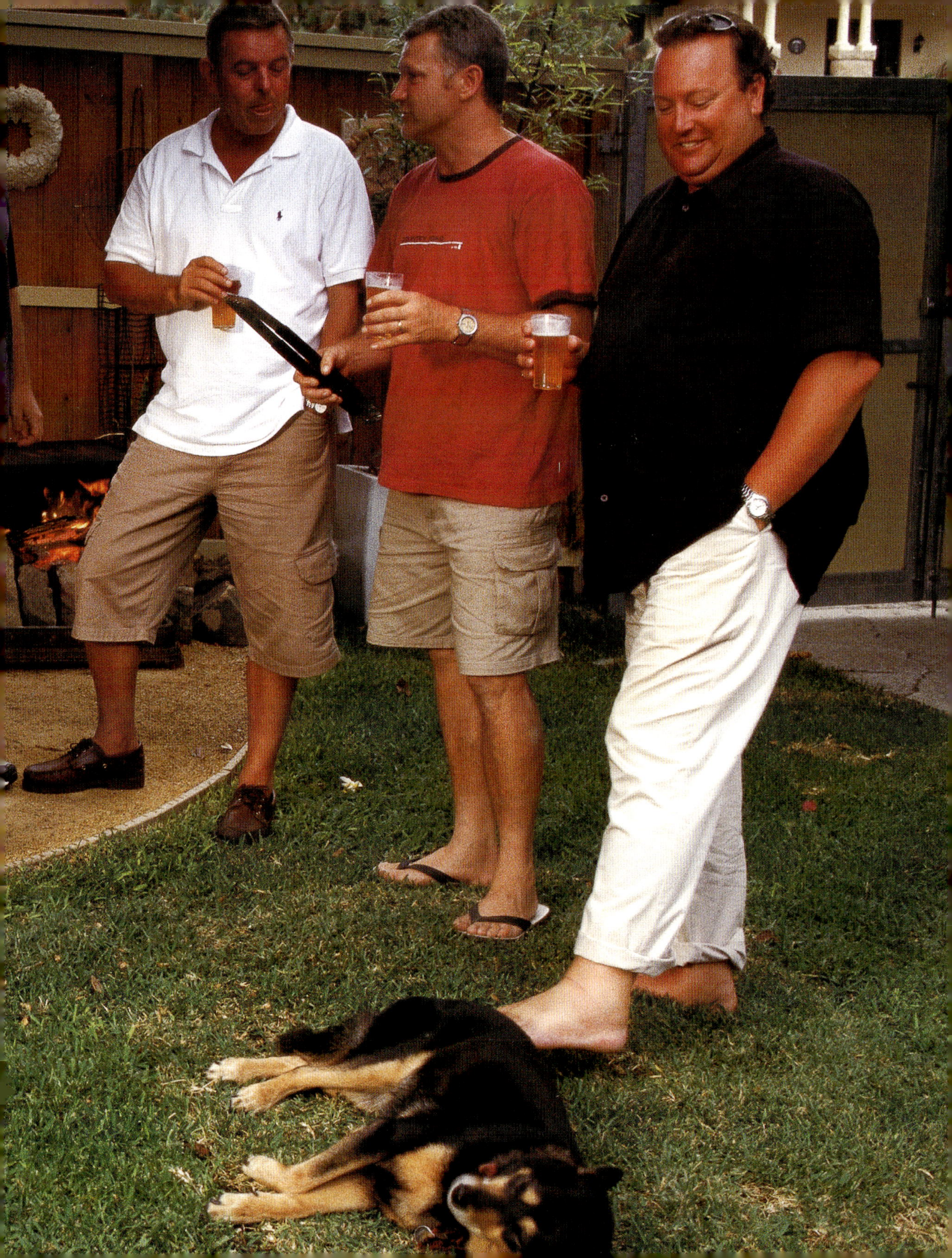

Recipes for knuckleheads

I don't mind jumping in the kitchen and having a cook-up. I tend to stick to traditional meals though, nothing fancy. The old go — rissoles or meatloaf (I'm still not up to scratch on Mum's meatloaf, but when you think about it, nothing comes close to my Mum's meatloaf).

Corned beef is my favourite meal — with white sauce, string beans and mash. I have it down pat, it's as good as Mum's...if not...No, I won't go there.

I have half a dozen mates who have never tasted corned beef. That's like not stopping for the Melbourne Cup. It's unAustralian. We had corned beef for tea once a fortnight when I was a kid. I could smell Mum's corned beef boiling away when I got off the bus after school, two streets away. I'll give you the traditional recipe to get you started, plus some favourite ones from the blokes on the Table of Knowledge.

My favourite meal

I WAS TALKING TO THE BOYS AT THE TABLE OF KNOWLEDGE ONE NIGHT, AND ZOLLY, WHO HAS A HUNGARIAN BACKGROUND, SAID HE HAD NEVER EATEN CORNED BEEF. I SAID, 'WHAT! WERE YA BORN ON MARS OR SOMETHING!' HIS MUM COOKED TRADITIONAL HUNGARIAN MEALS, AND CORNED BEEF WASN'T ONE OF

THEM. ZOLLY WAS MORE LIKELY TO GET GOULASH. AS I SAID, CORNED BEEF IS MY FAVOURITE MEAL, AND HERE IS HOW TO COOK IT. IT MAY NOT BE PERFECT, BUT THIS IS HOW I DO CORNED BEEF IN THE CAM HOUSEHOLD EVERY FORTNIGHT.

Whip down the butcher and grab a nice big piece of silverside, at least 1.5 kg (3 lb 5 oz). Get him to weigh it, as knowing how heavy it is will be important later. Buy more than you need because the second best feed in this world is cold corned beef sandwiches with heaps of salt.

The best feed of course is, this one I'm about to give you, and I think you'll agree that out of all the Table of Knowledge recipes, mine is the best.

It always turns out that with just about everything I cook, you get to stand back and have a few beers, just watching whatever it is cook. The good, old corned beef is no different. Similar to Zolly's five stubby veal goulash on page 184, this is a seven-course meal — one six-pack of cold cans and a big lump of corned beef.

Cam's corned beef

Ingredients
1 TABLESPOON VINEGAR
12 WHOLE BLACK PEPPERCORNS
2 WHOLE PEELED ONIONS
4 CARROTS, PEELED AND HALVED

Method

1 USING ONE OF THOSE POTS THAT HAS A HANDLE ON EACH SIDE AND A LID, DROP IN THE CORNED BEEF AND TOP UP WITH COLD WATER, MAKING SURE THAT ALL THE MEAT IS COVERED.

2 ADD THE VINEGAR, BLACK PEPPERCORNS, ONIONS AND CARROTS.

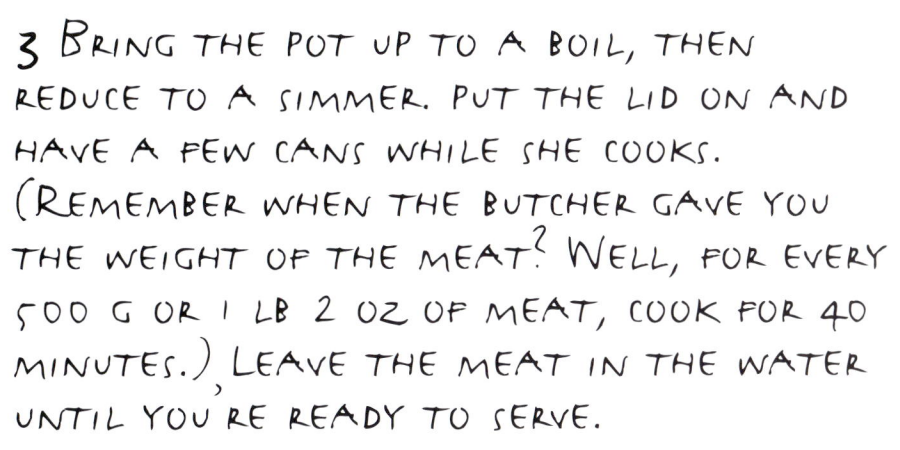

3 BRING THE POT UP TO A BOIL, THEN REDUCE TO A SIMMER. PUT THE LID ON AND HAVE A FEW CANS WHILE SHE COOKS. (REMEMBER WHEN THE BUTCHER GAVE YOU THE WEIGHT OF THE MEAT? WELL, FOR EVERY 500 G OR 1 LB 2 OZ OF MEAT, COOK FOR 40 MINUTES.) LEAVE THE MEAT IN THE WATER UNTIL YOU'RE READY TO SERVE.

WHEN IT'S READY, THE CARROTS AND ONION WILL BE THE BEST YOU'VE EVER EATEN. ADD some mash and string beans to the party, but the most important thing is to pour white sauce over everything (see the Boss's recipe below), then sprinkle over salt and cracked black pepper.

The Boss's white sauce

We're a big family, so the Boss makes us 2 cups of sauce.

Ingredients

60 g (2¼ oz) butter
½ cup flour
2¾ cup milk

Method

1 In a saucepan over a low heat, melt the butter.

2 Remove from the heat and add the same amount of flour, stirring until the mixture is smooth.

3 Now gradually stir in ¼ cup of milk. Stir it in until it's nice and smooth.

4 Return the sauce to the heat, constantly stirring. As it thickens up, remove from the heat and add the

remaining 2½ cups of milk until the sauce reaches the desired consistency. Simmer for about 10 minutes and season to taste.

Poly's scrambled eggs

This quick and simple recipe will delight the tastebuds. Serves one person.

Ingredients

2 eggs
100 mL (3½ fl oz) milk
⅓ cup halomi cheese, grated

Method

1 Crack the eggs into a saucepan. Beat the eggs, then slowly add the milk.

2 Grate the cheese into the mixture.

3 Place the saucepan on a slow heat. Gently fold the mixture with a wooden spoon until it slowly sets and becomes juicy. (Takes about nine minutes.)

4 Take the saucepan off the heat and let it sit for 30 seconds. Garnish with parsley and serve with your choice of sausages, bacon, tomato and mushrooms.

Zolly's five stubby veal goulash

You'll need five stubbies of beer here — just have one at regular intervals.

Ingredients

2½ tablespoons oil
1 small onion, finely diced
3 teaspoons mild paprika
500 g (1 lb 2 oz) veal, diced
2 tomatoes, diced
1 red capsicum, diced
1 teaspoon salt
1 small red chilli, seeded and diced
1 clove garlic, crushed

Method

1 Heat 1½ tablespoons of oil in a large saucepan over medium heat. Add the onion and cook it until golden, stirring from time to time.

2 Remove the saucepan from the heat. Add 1 tablespoon of oil, then stir in the paprika well.

3 Put the saucepan back on a low heat, add the veal and stir until it changes colour without browning (about 2–3 minutes).

4 Stir in the tomatoes and capsicum.

5 Add the salt, chilli and garlic, and then stir.

6 Cover, and cook until tender (about one hour).

7 Remove the lid, then let it simmer for about five minutes to thicken the stew. Add salt to taste, then serve with rice.

Diamonds

A few years ago, a couple of mates went out and bought their wives diamond rings (eternity rings, they're called). I reckon I should be getting something for the eternity bit, like a diamond the size of a tennis ball.

The Boss was seeing all these rocks floating around, and I could see the look on her face. But she never said a word, so I thought, 'Better get something special for her, maybe some diamond earrings.'

Eventually, I stumbled across this joint that sold diamonds. A bit of a bonus was that the bloke could set them for me. I told the jeweller that I was after some diamond earrings for the Missus.

'How much would you like to spend, sir?'

I said, 'About 500 bucks, mate.' He looked at me as if I had said Bradman couldn't bat. He unfolded the velvet and reached in with the tweezers. He brought the tweezers up and held them in the light. 'For $500 you'll get two of these in a gold setting.'

I looked at the tweezers. I thought I must be a bit far away. I got closer, then closer, until I was holding the tweezers an inch from my eye. I said to the bloke, 'You're having a lend of me, mate. There's nothing in there. The tips of the tweezers are touching.'

He said, 'Look closer.' And sure enough I saw it — but only for a split second. I was squeezing too hard on the tweezers, the tips made a little sound as they crossed over each other, and 'dink', the diamond was gone.

'Shit, I've dropped it!' The jeweller bloke and I were on our knees, looking for something about the size of a bee's ball. Then all of a sudden, incredibly, I saw the little sparkle under the desk. I licked my index finger, reached over, pushed down on it and up came the diamond — GOT IT!! Neither of us could believe it.

Anyway, he made the setting big to make the rocks look bigger. Didn't matter anyway — she loved them, and always wears them out on special nights.

Vinnie's tips on frying schnitzel

Despite opinion to the contrary, the secret to schnitzel is not in its preparation, but in its frying. Chicken, veal, fish, calamari or mushrooms — whatever you like, you can crumb it and fry it.

So pick up any decent recipe book to work out how to crumb.

Method

1 The basic routine is to coat the food in flour, then egg, then breadcrumbs. I add heaps of crap to the egg part, such as milk, lemon juice, parmesan cheese, parsley and pepper.

2 Try and use a good frypan. Frypans that disperse heat evenly are best. They also happen to be the most expensive. If you have a cheap frypan, you'll have to move the cooking food around too much.

3 Put heaps of oil in your frypan, at least 1 cm (1/2 in) deep. Do not be all

health-conscious about the oil. It's fried food, for God's sake! The more oil the better! I use vegetable oil; however, the guys at the pub reckon that sunflower and peanut oil are just as good. Olive oil is too expensive and also too strong in flavour.

4 Now the most important part of the whole thing: the oil must be at around the right temperature. The heat should be medium to high, but not too hot. The best way to test the heat is to put one of the crumbed pieces in. Listen. If it does not make a sizzling sound, it's not hot enough. If it sounds like it's really splattering, it's too hot. Cook each side for 2–3 minutes.

5 If you're cooking more than one batch, always wait for the oil to reheat. The more you put in at a time, the more the temperature of the oil will drop, making the food soggy. If the food is cooked correctly, you'll find that no oil will soak through the crumb mixture. It will also be crisp and golden on the outside and tender inside.

6 And finally, always drink and sing when you're cooking. Buon appetito!

INDEX

A
AGRICULTURE DRAINS, 59–62
AKUBRA HATS, 112
ANGLES
   3:4:5 METHOD, 17–18
   CUTTING, 78
ANN, 12, 48, 171–3, 186, 187

B
BACKYARD BARBIES, 123–30
BACKYARD MAINTENANCE, 84
BACKYARD PROJECTS, 14, 16
BAR ETIQUETTE, 159
BARBECUES **SEE** BARBIES
BARBERS, 12–13
BARBIES
   ETIQUETTE, 148–50
   GAS, 124
   PLATES/GRILLS, 143–4
   SEE ALSO OPEN-FIRE BARBIES; OPEN-FIRE COOKING
BARREL BOLTS, 105
BASEBALL CAPS, 112–13
BEAMS
   DECORATIVE, 43–5
   PUTTING UP ALONE, 47
BEARER HEIGHT, 30
BEER, 51, 52–3
   BAR ETIQUETTE, 159
   BOTTLE OPENERS, 156
   COOLING, 152, 155, 159
   DRAFT BEER AT HOME, 151, 155
   HOME BAR, 151–5
   KEG BEER, 152–5
BILL, 171
BLOKOSEXUALS, 10–13, 175
BOAT PURCHASE, 90–6
   BALMAIN TO ROSE BAY TRIP, 93–6
BOOTS, 97
BOTTLE OPENERS, 156
BRICKS, 38, 118
BUILDERS SQUARE, 103
BUSH BLOCK, 131–3

C
CAMP-OVEN COOKING, 134–41
   OPEN-FIRE PIT, 135–7
CAMPING, 11–12, 131–3
CAR MAINTENANCE, 114–18
   GREASE CHANGE, 116
   OIL CHANGE, 114–15, 116
   OIL FILTER, 116
CARMO, 107–11
CEMENTING IN, POSTS, 18–21
CHARCOAL, 125
CHARLIE, 8–9, 74, 131–3, 171
CHOOFERS, 166–70
CLEANSING CREAM, 173
COLD NIGHT SLEEPOUT, 132–3
COMPRESSORS, 101, 164–5
CONCRETE
   NORMAL, 21
   QUICK-SETTING, 20–1
CONCRETE BAGS, 20
CONDOBOLIN, 108
CORNED BEEF, 178–82

D
DAMPER, 147
DECKS, 26–34
   BEARERS, 29–31
   DECKING BOARDS, 32–4
   DECKING NAILS, 32
   JOISTS, 31–2
   SET-OUT, 26–8
   WOOD PRESERVATIVE, 34
DEHYDRATION, 113
DIAMONDS, 186–7
DOG CARE, 80–3
DOG KENNEL, 75–9
DOMESTIC DUTIES, 163
DONKEYS, 166–70
DRAFT BEER, 151, 155
DUSTING, 164–5

E
EARRINGS, 186
EGGS, SCRAMBLED, 183
ETERNITY RINGS, 186

F
FACIAL HAIR, 174–5
FACIALS, 9, 12
FENCING, 98–101
FENCING GUN, 101
FIRE BANS, 130, 141
FIRE BUILDING, 9, 126–30

FIRE SAFETY, 141
FIRES, EXTINGUISHING, 130
FITTINGS, GALVANIZED, 22
FLANNO SHIRTS, 120-1
FOLDING GRILL, 126

G
GALVANIZED FITTINGS, 22
GARDEN ISLAND, 95-6
GATES, 102-5
GEO-TEXTILE FABRIC, 61-2
   SOCK, 62
GRAPEVINES, 46-7
GREASE CHANGE, 116
GREASE NIPPLE, 116
GREENPEACE, 95
GRILL, FOLDING, 126
GRIP DECKS, 33

H
HAIR CUTS, 12-13
HARDWOODS, 35
HATS, 112-13
HOME BAR, 151-5
HOOP IRON, PERFORATED, 72-3
HOT CLIMATE
   FLUIDS, 113
   HATS, 112-13
HOT WATER SYSTEM, HOMEMADE, 166-70
HYPOTENUSE, 17-18

I
ICE, 159

J
JANZ, GEOFF, 148
JOIST HANGERS, 45
JOISTS, SKEW-NAILING, 68

K
KEG BEER, 152-5
KENNEL, 75-9
KEV, 161
KINDLING, 128-9

L
LADDER, TO TREEHOUSE, 74
LAMB ROAST, IN CAMP OVEN, 134-41
LETTUCE, 10
LINE LEVELS, 17
LIZZIE, 80-3, 94

M
MAGIC BOX (BEER CHILLER), 152-3
MASSAGE, 172-3
MEATLOAF, 178
MOORING, AT ROSE BAY, 92-3
MOUSTACHES, 174-5
MUDGEE, 131, 132
MULCHING, 119

N
NECK MASSAGE, 172-3
NICKNAMES, 160-3
NORTHERN TERRITORY, 112

O
OIL CHANGE, 114-15, 116
OIL FILTER CHANGE, 116
OPEN-FIRE BARBIES, 12, 124, 142
   BRICK, 146
   THREE-STONE, 144-6
   TWO-STONE, 142-4
OPEN-FIRE COOKING
   FIRE BUILDING, 126-30
   FOLDING GRILL, 126
   KINDLING, 128-9
   RED-HOT COALS, 125
   WET TIMBER, 126
OPEN-FIRE PIT, 135-7
OREGON, 35-6
   DRESSED, 36
OUTBACK TREK, 106-11

P
PEGBOARD, WITH HAND TOOLS, 87-8
PERFORATED HOOP IRON, 72-3
PERGOLAS, 40-7
   DECORATIVE BEAM, 43-5
   RAFTERS, 45-6
   TIMBER SPANS, 40
PEWTER MUGS, 157-8
   PRIMING, 158
PICKET GATES, 102-5
PINE, TREATED, 21-3, 130
PINK-PRIMED TIMBER, 22
PLATFORM TREEHOUSE, 64-70
PLYWOOD, MAKER'S EDGE, 77
POOCH PALACE, 75-9

porkies, 87–8
post hole shovel, 18
posts, cementing in, 18–21
pulley, 64, 69–70

R
retaining walls, 54–8
   agriculture drains, 59–62
right angles, 3:4:5
   method, 17–18
ripple face, on boards, 33
road rage, 39
Royal Flying Doctor Service (RFDS), 106
   Outback Trek, 106–11

S
sandpit, 70
sandstone, collecting, 145
Sarah, 171
sausages *see* snags
schnitzel, 188–9
scissor shovel, 18
scrambled eggs, 183
set-out, 26–8
shaving, 174, 175
shed, 9, 86–9
   equipping, 86, 87, 89
sign writing, 38
skew-nailing, 68
sleep clinic, 48–53
sliding bevel, 78
snags
   cooked by grandfather, 124–5
   cooked to perfection, 150
SNAGS (Sensitive New Age Guys), 8, 16
snoring, 48–50, 53
sport, 10–11
string lines, 98
   setting up, 16–18
stubby holders, 121
sump, 114
sunstroke, 113

T
'The Castle', 90, 131
three-stone open fire, 144–6
timber, 21–3
   cut ends, 23
   hazard rating, 21–2
   pink-primed, 22
   spans, 28, 40
   treated, 21–3
'Trading Post', 90
treated pine, 21–3
   cut ends, 23
   nailing, 33–4
   not for burning, 130
treehouses, 63–74
   ladder to, 74
   platform treehouse, 64–70
   sandpit under, 70
   true treehouse, 71–4
turf removal, for fire, 136
turf replacing, after fire, 141

U
USS Constellation, 95–6
utes, 37–9
   with signs, 38

V
veal goulash, 184–5
vehicles *see* utes
Vinnie, 174, 188

W
wall plates, 42
walls, retaining, 54–8
water level (tool), 23–5
water use, 119
weeding, 119
white sauce, 182–3
work boots, 97

Z
Zolly, 179, 184

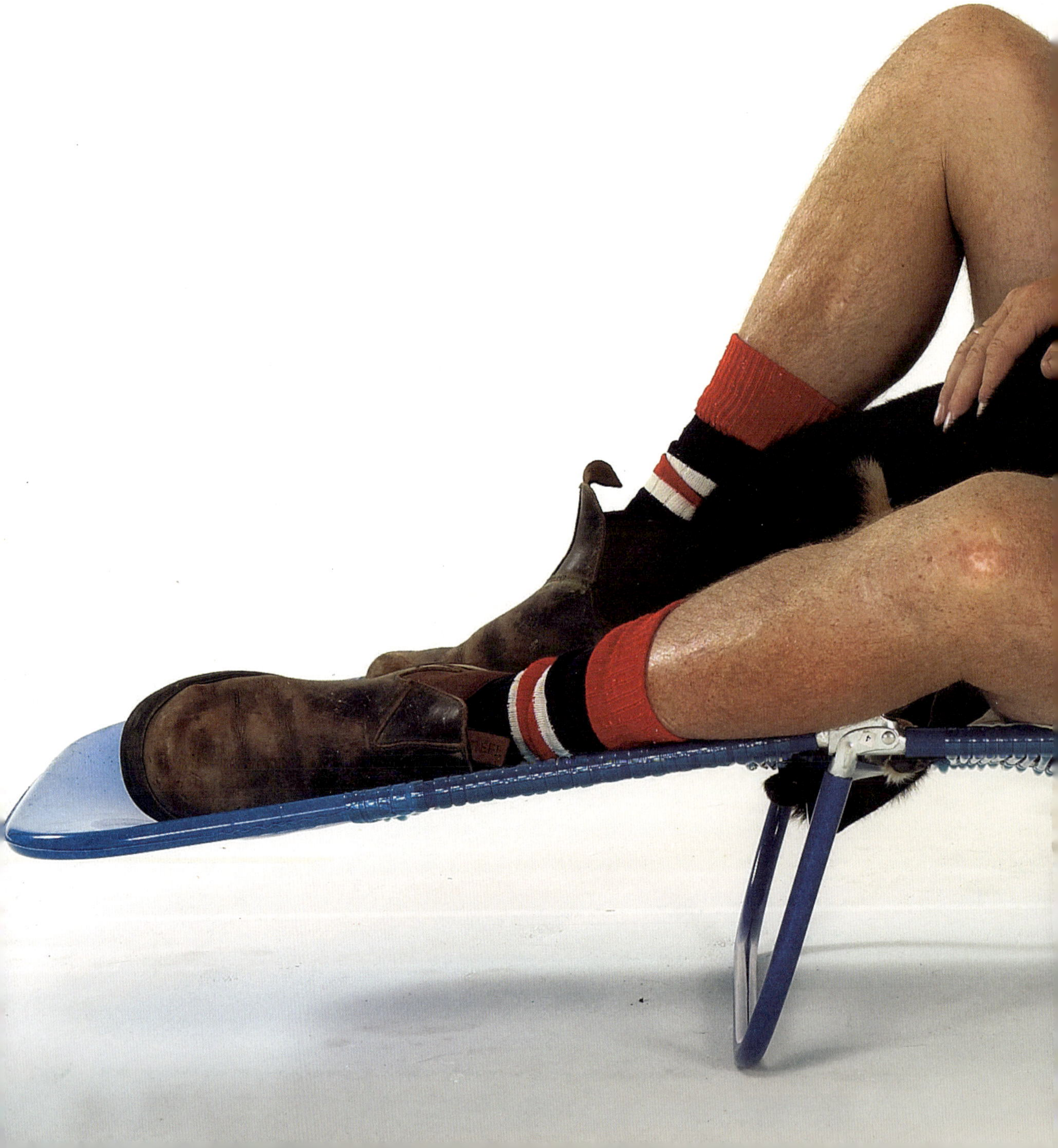